The Works of Isambard Kingdom Brunel

Isambard Kingdom Brunel
from the engraving by T. O. Barlow

The Works of Isambard Kingdom Brunel

AN ENGINEERING APPRECIATION

Edited by
SIR ALFRED PUGSLEY
Emeritus Professor of Civil Engineering
University of Bristol

CAMBRIDGE UNIVERSITY PRESS

Cambridge

London New York New Rochelle

Melbourne Sydney

CAMBRIDGE UNIVERSITY PRESS
Cambridge, New York, Melbourne, Madrid, Cape Town, Singapore,
São Paulo, Delhi, Dubai, Tokyo, Mexico City

Cambridge University Press
The Edinburgh Building, Cambridge CB2 8RU, UK

Published in the United States of America by Cambridge University Press, New York

www.cambridge.org
Information on this title: www.cambridge.org/9780521157780

First published jointly by the Institution of Civil Engineers and
the University of Bristol 1976

First paperback edition 2010

A catalogue record for this publication is available from the British Library

British Library Cataloguing in Publication Data
The works of Isambard Kingdom Brunei
1. Brunel, Isambard Kingdom
2. Civil engineering
I. Pugsley, Sir Alfred
624 TA140.B75 79-41470

ISBN 978-0-521-23239-5 Hardback
ISBN 978-0-521-15778-0 Paperback

Contents

continued

Figures

continued

Acknowledgements

This book makes much use of the collection of Brunel's papers in the Library of the University of Bristol, and the authors wish to thank the University Librarian and his staff, particularly Mr G. E. Maby, for facilitating this. These papers comprise a series of letter books, calculation books and sketch-books presented to the Library by the late Celia, Lady Noble of Bath, a grand-daughter of Brunel; and also the minute books and other records of the Clifton Suspension Bridge Company, presented by the Bridge Trustees.

We are indebted also to Sir Marc Brunel Noble, a grandson of Lady Celia, for arranging for us to consult Brunel's diaries and other engineering papers in his possession; and to the Chief Civil Engineer (Western Region) of British Rail, for giving us access to the many fine engineering drawings relating to the Great Western Railway.

We have to acknowledge the generous help afforded by the Librarians of the Institution of Civil Engineers and of the Royal Institution of Naval Architects.

No book of this sort could have been produced without the background provided by the biographies of Brunel that have been published; in particular we would acknowledge the help afforded by those by I. Brunel and by L. T. C. Rolt.

Introduction

This book has its origins in the two institutions that have come together as its joint publishers. The University of Bristol, close to the scene of much of Brunel's early work, and custodian of so many interesting Brunel papers, could hardly do other than play a part, even had its present Vice-Chancellor, Sir Alexander Merrison, at the suggestion of Professor R. L. Gregory of its Publication Committee, not been the first to initiate some such publication. The Institution of Civil Engineers, in whose affairs Brunel played such a lively part for the whole of his working life that he was about to become its President when he died, and whose then senior members were drawn to complete "his first love"—the Clifton Suspension Bridge—in his memory, naturally entered into the partnership with enthusiasm.

Both bodies have deeper connections with Brunel and his works, and with each other, than the foregoing. Engineering in the University derives from two main streams of relevant educational activity in Bristol: one centred in University College, to the celebration of whose centenary in 1976 this book contributes, and the other at the Merchant Venturers' Technical College established in 1893 but having its origins in a pioneer "trade" school founded in 1856. Both were active in engineering education and came together as a University in 1909. At that stage R. M. Ferrier became the first Professor of Civil Engineering and the University Engineering Board was specially strengthened by the appointment to it of Sir James Inglis, then President of the Institution of Civil Engineers and General Manager of the Great Western Railway. Among the University's early undergraduates was A. J. S. Pippard, who later held the University Chair in Civil Engineering and became President of the Institution in 1958. As such he took part, together with Lady Noble, Brunel's granddaughter, in commemorating at Clifton in 1959 the centenary of Brunel's death.

But the University has a more intimate but little known connection with Brunel. When the University was founded, a Vice-Chancellor with experience in the organisation of a new university was sought, and was found in Sir Isambard Owen, who had just helped to establish the University of Wales. He was the son of W. G. Owen, a railway engineer who had worked under Brunel on the railways of South Wales and who later himself became Chief Engineer of the Great Western Railway. It was natural, therefore, that Owen's son should have Brunel as his godfather and be given the name of Isambard.

The book that has grown out of the interest of these two institutions sets out to discuss Brunel's works in engineering terms, but it has of course proved impracticable to consider his every engineering activity. Nothing has been said, for example, of his early work with his father on carbonic acid gas as a source of motive power or of his contributions to dock engineering. But all his major works have been treated in some detail by appropriate engineering specialists, and it is hoped that by their so doing engineers and others will find his skills and limitations highlighted in a way that has not been done before.

Two chapters—the first and the last—differ somewhat in character from the rest. The first relates to Brunel and his engineering staff. It is a curious fact that, while many biographies of Brunel have been written, none has had much to say about his relations with his staff. Yet every engineer knows that however hard he works—and Brunel certainly worked hard—no substantial engineering work can be constructed without the help of other engineers.

Engineering is, as the original Charter of the Institution emphasised, both an art and a science, and any assessment of the work of a great engineer must include some consideration of its scientific aspect and in particular, in Brunel's case, its mathematical aspect. So the last chapter of this book looks at this vital background to Brunel's work and his contributions to it.

Brunel grew up and worked during a unique period of British civil engineering that lasted for nearly a century. It can be said to have started with John Smeaton, Thomas Telford and John Rennie—a triumvirate working mainly with masonry and roads—and to have ended with a second triumvirate, even more closely related—Robert Stephenson, Joseph Locke and Brunel—working mainly with wrought iron and railways. Together these men and their leading contemporaries placed British civil engineering on a world peak that has not been achieved since.

Contemporary conditions were such as to foster their works; the industrial revolution, which encompassed much more than civil engineering alone, was under way, and the country was reacting to the Napoleonic threat. Indeed, the conditions at the start of the nineteenth century, that gave birth to modern civil engineering, had much in common with those at the start of the twentieth century, when this country was both experiencing the stirrings of a scientific revolution and reacting to the German threat. As a result a new band of engineer-scientists emerged to create aeronautical engineering and place it on a world peak that saved their country.

The great British engineers of the first half of the nineteenth century did more than build roads and bridges and railways: civil engineering in their hands became a profession and was raised to a new social level. It was not long after the establishment of the Institution of Civil Engineers in 1818 that many of its leading members were elected to the Royal Society. Among these Sir Marc Brunel[1] was one of the first, and the "second triumvirate", to which his son belonged, were all elected by 1850. It was natural to find that nearly all the large engineering contingent that thus entered the Royal Society were elected also to the Athenaeum, of which the Brunels and Robert Stephenson became particularly popular members.

An interesting reflection of these developments is to be seen in the country homes that these pioneering professional engineers acquired or built for themselves as they approached retirement. With increasing prosperity, their homes mounted in size, from George Rennie's "Priory" near Newbury in Berkshire to Sir John Fowler's fine estate of 40,000 acres at Braemore in Ross-shire. Thus Brunel was not unusual among his engineering contemporaries in making his London home in Duke Street, Westminster—then fashionable enough for his even more ambitious contemporary, the young Disraeli—and in planning in his last years for a country home at Watcombe, near Torquay, an area he had learned to love during his railway work in South Devon. Although the large house he sketched was never built, Brunel and his aged father, then confined to a wheel chair, spent many happy hours viewing the site and planting trees and shrubs to embellish the grounds.

This book naturally provides material for a reassessment of Brunel as an engineer.[2] Each chapter presents a picture of his skill and his shortcomings, his boldness and his energy. From the start it is clear that he combined a readiness to learn from others with a flair for bold experiment based on his own convictions. In his willingness to study the work of others he followed his father; both had the same observant nature, even in small matters. Thus we find Brunel noting, no doubt for comparative purposes in the future, the inclination of Park Street, a well-known steep hill in Bristol, while Sir Marc breaks off in his diary to note the dimensions of the rise and tread of the grand staircase in the Athenaeum! In his bias for bold experiment, Brunel went further than his father, although the latter was mechanically the more inventive. On these matters the obituary written for the Royal Society at Brunel's death still has much relevance. After noting that "Mr. Brunel was fervently attached to scientific inquiry" the writer concludes:

> "His carelessness of contemporary public opinion, and his self-reliance, founded on his known character and his actual works, was carried to a fault." (Royal Society, 1860)

Brunel's standing *vis-a-vis* his father and his greatest contemporary, Robert Stephenson, is of special interest. In England, Brunel's contemporaries would certainly have placed all three in the "top ten" and probably put Brunel and Stephenson in the "top five". On the Continent, however, Brunel was not (and still is not) so appreciated. His father came first there, to be passed later by Stephenson. The Brunel family, represented now by the Nobles, has, until recent decades, looked upon Sir Marc as an engineering "success" and his son as a "glorious failure"; and Lady Noble, Brunel's grand-daughter, was surprised by the public interest and enthusiasm evoked by Rolt's book, *Isambard Kingdom Brunel*, when it was published in 1957.

Civil engineers of today clearly regard Brunel as one of the greatest of Victorian engineers, and the authors of this book have found it a privilege as well as a labour of love to examine his engineering works.

SIR ALFRED PUGSLEY

I

I. K. Brunel: Engineer

DR R. A. BUCHANAN

It in no way detracts from the achievements of Isambard Kingdom Brunel to see him as a man of his times, who had the ability to take advantage of the opportunities offered by those times, before the engineering profession had specialised into civil, mechanical, marine, electrical and other sections which constricted the imaginative possibilities of any one engineer, however able he might be. Brunel was a polymath who had the good fortune to flourish in the few brief decades when society needed the services of many-sided engineers. These were the early Victorian decades from 1830 to 1860—a period of rapid social and political change, industrial and commercial expansion, and of unquestioned world leadership for Britain. The reformed House of Commons after 1832 brought many people of Brunel's social class to political power for the first time, and although Brunel did not succumb to the temptation of standing for Parliament himself, like other great engineers of the period including Robert Stephenson, his work as engineer for many companies seeking statutory powers ensured his intimate familiarity with the corridors of political power. In the Chartist panic of 1848 he enrolled as a Special Constable, and when the Crimean War revealed the serious military and administrative inadequacies of the British forces Brunel was ready to help with plans for floating gun batteries and prefabricated hospitals. He was thus an eminent early Victorian, imbued with the doctrines of competitive freedom, *laissez-faire* and self-help, even though Samuel Smiles, whose best-selling work *Self-Help* was published in the year of Brunel's death, did not include him in his pantheon of great engineers. He carried his insistence on *laissez-faire* to the extent of rejecting patent protection, which he never sought for any of his own innovations, and his relationships with his own assistants tended to be authoritarian and paternalistic in the tradition of the early Victorian family. It is too easy to see Brunel as a man born before his time, in that many of his projects, including the atmospheric railway, were in advance of technical skill to execute them or, like the steamship *Great Eastern*, came before there was an economic function for them to perform. He was undoubtedly a man of vision, even of genius. But he was always firmly rooted in his own period, and it is as a great Victorian that he should be seen.

Brunel's first professional engagement as a civil engineer was as Resident Engineer for his father on the Thames Tunnel, but this project was put in abeyance for many years after the accident in 1828, from which the young Brunel narrowly escaped with his life, when the Thames broke in and flooded the workings. The first commission he received in his own right was that for the construction of the Clifton Suspension Bridge, following the success of his entry for the competition to choose a satisfactory design, and the link thus established with Bristol business circles led to other invitations such as that in 1831 to report on the condition of the Bristol Docks and that in 1833 to become Engineer for the London to Bristol railway project, shortly to become the Great Western Railway.[3] The launching of this enterprise marked Brunel's establishment as a famous engineer whose talents were much in demand for all sorts of projects. Musing on his successes of the past year on Boxing Day 1835, Brunel observed that while still only aged 29 he had been entrusted with over £5 million of capital expenditure (Rolt, 1957). He was then about to move into his new house at 18 Duke Street (convenient for Parliament and Whitehall), which remained his home and his office for the rest of his life. From here he administered the army of assistants, resident engineers and pupils which he assembled to undertake the many complicated engineering operations in which he became involved. Written communications to these subordinates represent a high proportion of the entries in the series of private letter books which, fifteen in number, began about this time.[4] Even though the series of private letter books, now in the Bristol University Library, covers only part of Brunel's career, and omits except incidentally the business of the Great Western Railway, the Great Western Steamship Company and the enterprises involved in the construction of the *Great Eastern*, all of which probably had letter books of their own, it contains material which is of crucial importance in arriving at an understanding of Brunel's self-consciousness of himself as an engineer. The quality of this self-consciousness is elucidated here mainly by drawing on the private letter books.[5]

By the beginning of 1836, Brunel already had a staff of considerable, although indeterminate, size. Beginning with W. H. Townshend, a Bristol land surveyor who had initially been in charge of construction work on the Bristol and Gloucestershire Railway, and who had been a potential rival to Brunel as Engineer to the GWR but who agreed to act as his assistant surveyor, he had quickly acquired the help of assistant engineers, surveyors and contractors on his railway projects. Among the first of these were William Gravatt, who had worked with the Brunels on the Thames Tunnel and who was responsible for surveying the Bristol and Exeter Railway, and J. W. Hammond, who became his most trusted assistant on the GWR works and his chief assistant until his death in 1847. Each major project required at least one Resident Engineer and sometimes also an assistant engineer and a surveyor, and as in each year from 1836 Brunel had something like a dozen such projects (he listed eleven in the journal account of Boxing Day 1835) his staff must regularly have been between 30 and 40 strong, not including pupils, local services employed on his behalf by his assistants, contractors and their workers, or members of important subsidiary staffs such as those of the GWR engineering workshops under Daniel Gooch. Because Brunel conducted so many of his transactions with his staff by face-to-face contact in the course of his extensive travelling away from Duke Street, the letter books are an inadequate base from which to form a precise picture of this intricate team, but we are fortunate that on one occasion Brunel sent a circular letter (private letter book, *PLB, 5–6 Mar., 1850*) to all members of his staff which is recorded. This was signed by his chief clerk Joseph Bennett and refers to a notice for a competition, presumably connected with preparations for the Great Exhibition a year later:

> "Dear Sir,
>
> "Mr. Brunel calls your attention to the enclosed and hopes you will try and send some good design . . ."

There follows a list of 33 names to whom the letter was addressed: R. P. Brereton, T. A. Bertram, R. Brodie, T. E. Blackwell, B. A. Babbage, A. J. Dodson, G. J. Dailey, C. E. Gainsford, W. Glennie, J. B. Hannaford, J. Hewitt, R. W. Jones, Saml. Jones, M. Lane, E. F. Murray, P. J. Margary, Saml. Powers, Wm. Peniston, W. G. Owen, C. Richardson, H. Savage, C. Turner, R. Varden, W. Warcup, R. J. Ward, P. P. Baly, R. Beamish, W. Bell, O. C. Edwards, L. C. Fripp, J. Gibson, J. H. Hainson and G. F. Okeden. All were addressed as

"Esq." which indicates that they were all regarded as gentlemen; otherwise, observing Victorian conventions for addressing the lower orders, Brunel would have used the surname only and dropped the prefix "Dear" at the head of the letter.[6] The letter was written after Hammond's death, when R. P. Brereton had become his chief assistant, a post which he retained until Brunel's death. With Bennett, Brereton had the task of sorting out Brunel's affairs after his death and assumed responsibility for a number of unfinished projects. We will meet again several of the names in this list, but for the moment our purpose is only to get some impression of the scale of Brunel's engineering enterprise. March 1850 may not have been a particularly good time in Brunel's fortunes, as there are several indications that in the previous year he was experiencing some economic anxiety:

> "These are *very* economical times. I foresee an extreme probability of the GWR Directors calling upon me to make an almost clean sweep of all expenses on the Wilts & Somerset . . . no body must *rely* upon employment after this quarter." *(PLB, 4 May, 1849)*

On the other hand, there were dismissals shortly after the letter went out, as J. B. Hannaford and five other members of the drawing office staff who do not appear in the previous list were laid off later in the same month *(PLB, 25 Mar., 1850)*. So our estimate of the average size of Brunel's senior staff is almost certainly of the right order of magnitude.

With such a large and constantly changing body of assistants, it was inevitable that Brunel should have had some discipline troubles, but the surprising thing is that they were not greater and Brunel's admirers are almost certainly correct in claiming that he had great skill in selecting the right sort of people to help in his work. The top people in the team, Bennett and Brereton, had the complete confidence of their chief. Brunel was clearly writing with some satisfaction in his own choice of Bennett when he advised the promoters of the incipient South Devon Railway in 1836 that success would depend most on the character of the secretary:

> ". . . nothing can be more erroneous that to suppose that a clerk is all that is required—a Secretary must in fact be able to assist the Directors in their decisions and must be able to command such confidence on the part of the Directors in his judgment and opinions as will allow him to act frequently upon his own responsibility—I should go as far as to say that an inefficient Secretary might be more injurious to a Company even than an inefficient Engineer." *(PLB, 28 Mar., 1836)*

Brunel never had any cause to regret his trust in Bennett. Similarly

The Brunel Society

Fig. 1. Robert Pearson Brereton

with Robert Pearson Brereton (Fig. 1), who joined Brunel's staff in 1836. He was sent out to Italy in 1845 to help Brunel to sort out some embarrassing complications which had arisen in the construction of the Genoa to Turin railway, and was described by Brunel as "my assistant, a peculiarly energetic persevering young man".[7] J. W. Hammond, whom Brereton succeeded as chief assistant, did have occasion to receive some less than friendly remarks from Brunel:

> "When I referred to our 'angry discussion' I meant simply and truly that I was angry, and if I failed to make you sensible that I was angry I must be a much milder and gentler being than I thought." *(PLB, 27 Nov., 1846)*

The point at issue was Hammond's slowness in fulfilling Brunel's instructions to despatch some railway waggons down to South Devon. But as Hammond was then acting for Brunel on the difficult and protracted labour of reconstructing the South Entrance Lock to Bristol Docks, there may have been other causes of irritation, and as Hammond died the following year it is possible also that his reactions had already slowed below the 100 per cent efficiency required by his chief.

The disciplinary cases which appear in the letter books make interesting reading, because they reveal a lot about Brunel's methods of handling his staff. One of the earliest to be recorded concerned Gravatt, who had done the preliminary work on the Bristol and Exeter Railway but who then appears to have quarrelled with Brunel, who wrote a long letter to him at the end of 1839, upbraiding him for "a most unprofessional act, sacrificing your duty to the Compy to me and to yourself, entirely to feelings".[8] By the middle of the following year the position had deteriorated, as Brunel wrote a curt note accusing his assistant of betraying him.[9] Whatever the cause of the trouble, the Directors of the B&E expressed their confidence in Brunel, who suggested that Gravatt be given a much more limited role in the works *(PLB, 4 Aug., 1840)*. The row blew up again the following year. Brunel wrote to his assistant regarding a B&E bridge near the Bristol terminus:

> "How could you leave me uninformed of the deplorable state of the bridge near the New Cut." *(PLB, 4 June, 1841)*

And the following week he asked Gravatt to resign.[10] Gravatt refused, and made countercharges which the Board decided to investigate, much to Brunel's annoyance.[11] The episode may have contributed to a souring of Brunel's relationship with the B&E Directors, and it is in a sense a minor precursor of the much more bitter struggle which

Brunel had later with John Scott Russell. But the letter books are silent on the immediate results of the affair.

About the same time another assistant, J. H. Gandell, came under suspicion of having been involved in some speculation at the temporary station at Farringdon. Brunel wrote *(PLB, 13 Jan., 1840)* demanding that he choose between his profession and the business of the speculator, in which "you will make *more money* and run less risk of *losing credit*". Gandell did not improve his ways to Brunel's satisfaction, particularly when it emerged that he had been suspected of misappropriating company supplies, on which charges his explanation failed to satisfy Brunel *(PLB, 4, 16 Mar., 1840)*. Gandell had been supplying a contractor improperly with materials, and was dismissed:

> "I do not consider that you have any claim against the Company for salary . . . I have no hesitation in saying that I think you were let off lightly much more so than you would had you been left to me alone." *(PLB, 2 Nov., 1840)*

Brunel required a high standard of gentlemanly conduct from his assistants:

> "I should look much more for these qualifications (ie. those required to serve the Great Western Steamship Company) in a moderate degree in a gentlemanly and trustworthy young man than for any scientific or ingenious person who might be disposed to act too much on his own responsibility." *(PLB, 13 May, 1846)*

He expected them to be courteous and tactful towards proprietors and directors, firm and distant towards contractors, fair and just towards subordinates, and he did not hesitate to reprove assistants who fell below these standards. One assistant to feel his lash on these points was R. M. Marchant, who had taken a high-handed attitude towards a subordinate called Hulme. Brunel defended the latter as "by Birth, Education and feeling a Gentleman in every sense of the word", and went on to criticise Marchant:

> "When a man complains of want of courtesy he should himself be most gentlemanly and courteous in his language which you are *very far* from being in your note to me . . ."[12]

On the other hand. Brunel was very conscious of the abuses of gentlemanly conduct. He explained to Robert Bird in 1841 that:

> "A short time back after repeated warnings to your grandson I was compelled to dismiss him from the Company's service as his excessive idleness not only rendered him useless but infected others." *(PLB, 3 June, 1841)*.

And in the same vein, some years later, he wrote *(PLB, 29 Dec., 1849)* to W. G. Owen, his Resident Engineer on the Chepstow Bridge, to enquire about a junior assistant, C. Smith, who had requested a testimonial:

> "I have an impression—if it is wrong correct me—that he is one of those who gets *up late*, go to their work at *gentlemanly hours*—and from whom it is difficult to get any real *work*."

It was clearly possible to be too much of a gentleman to suit Brunel, but in most respects he placed great emphasis on the qualities of integrity, reliability and courtesy. He had a particular dislike of vagueness and imprecision. Writing to an assistant in 1857 he said:

> "You are getting very loose I wont say careless but very *imprecise*—when it is a question of calculation . . . Your reasoning is (excuse my so describing it) Irish—it is loose and imprecise to a degree that makes it untrue."[13]

Another temptation of gentlemanly life against which Brunel kept a watchful eye was the appeal of sporting activity. He delivered a strong reprimand to an assistant in 1853 following some remarks from the Directors:

> ". . . as to an apparent want of energy and activity on your part in attending to the Company's works . . . contrasted with an alleged devotion to amusement and amongst other things cricket . . . I don't know why you should be less of a slave to work than I am, or Mr. Brereton, or any of my assistants in town—it would rather astonish anybody if Mr. Bennett should be a frequenter of Lord's cricket ground or practice billiards in the day time—and I don't know why a man having the advantages of country air and very light work should indulge them . . . You must endeavour to remove any such grounds of observation." *(PLB, 14 Sept., 1853)*

But the assistant so admonished managed to repair his damaged reputation, for exactly five years later Brunel invited him to be his Resident Engineer on the Bristol and South Wales Junction Railway, holding out the prospect of good cricket in the Bristol area:

> "I want a man acquainted with tunnelling and who will with a moderate amount of inspecting assistance look after the Tunnel with his *own eyes*—for I am beginning to be sick of Inspectors who see nothing—and resident engineers who reside at home . . . the country immediately north of Bristol I should think a delightful one to live in—beautiful country—good society near Bristol and Clifton etc.—I can't vouch for any cricketing but I should think it highly probable . . ."[14]

Nor had Brunel any objection to sport or recreation at the proper time and place. He defended one of his assistants vigorously against the charge of some Directors that he kept boxing gloves in his room:

> "I confess if any man had taken upon himself to remark on my having gone to the Pantomime which I always do at Xmas no respect for Directors or any other officer would have restrained me. I will do my best to keep my team in order but I cannot do it if my Master sits by me and amuses himself in touching them up with the whip." *(PLB, 19 Jan., 1842)*

On occasion, Brunel's defence of his assistants became almost a censure of those who ventured to express criticism. In speaking out for one Andrew of the "locomotive department" he admits that the man had peculiarities:

> "I have a high opinion of Andrew's integrity, good intentions, and abilities but his temper is the most singular I have had to deal with."

But he carries the attack into the camp of the complainant, who was J. Crosthwaite, a vocal member of the Liverpool faction which held an important group of GWR shares:

> "Upon my word you are the perfection of a contented proprietor—your property is at 75 per cent premium—and the higher it gets the more you grumble—I begin to suspect that like a horse's bite it is a proof of love." *(PLB, 30 July, 1844)*

Brunel had earlier had animated exchanges with Crosthwaite, as when he sought to demonstrate that the GWR had been economically built: "we look sharper after money than you suppose—though perhaps we dont talk so much about it down here as you do up in the north" *(PLB, 20 July, 1842)*.

It is impossible to reconstruct the relationship between Brunel and all his assistants, but a few provide interesting examples of how he treated them and of how they responded to his treatment. W. Froude had been in his service for some time in 1844, when illness obliged him to ask for leave. Brunel replied that it would be very inconvenient for the Company but instructed him to take off six months during which he must make a resolution "as you would a religious vow—if you ever make one" to have nothing to do with business "except such as *I* may trouble you with."[15] The therapy must have worked, for a year later Brunel offered Froude the post of Resident Engineer on the North Devon Railway.[16] Froude accepted, and Brunel wrote gratefully:

> "I have been compelled to the destruction of my comfort to undertake a great deal more than I can possibly attend to with credit or satisfaction to myself—and but for your help on the North Devon I should have dropped a huge stitch in my work." *(PLB, 26 July, 1845)*

Unfortunately, Froude was responsible for a major breakdown in the North Devon project, and offered to forego his salary. Brunel refused to accept this arrangement:

> "You did your best and the utmost I can say now that I am no longer afraid of annoying you is that you made a great mistake in not perceiving the danger sooner—quite a strange unaccountable mistake but from that very circumstance it is one of those which no one could impute to anything but a very singular accident . . . You must have the goodness therefore to send me your statement for your own salary." *(PLB, 19 Dec., 1845)*

Less than a year later Brunel was inviting Froude to help with another West Country line, for which his assistant Johnson would do the bulk of the engineering:

> ". . . but I want above that again a general who can judge of local interests etc. etc. policy etc. . . . It is I am sorry to say *not by any means* in conjunction with the B&E who having elected to be at war with the GWR have accepted my resignation as their Engineer . . ." *(PLB, 3 Oct., 1846)*

It is clear that Brunel valued Froude's powers of diplomacy as much as his engineering skill. The Johnson to whom he refers was W. Johnson, an elderly but very competent surveyor who had already done good work for Brunel on the Sardinian Railway project *(PLB, 20 July, 1842)* and on a survey in Northumberland *(PLB, 8 Nov., 1844)*. Brunel tried to lure Johnson out of retirement in 1853 to undertake another Italian railway survey for him *(PLB, 14 Nov., 1853)*. It is not clear whether or not Froude accepted this particular invitation, but he certainly fulfilled other commissions for Brunel, including some "theoretical experiments":

> "I deny altogether the very foundation of your theory now that you lay it bare—the rate of expansion will *not* be infinite in the case you assume . . . I feel that I may be altogether writing nonsense—one sadly loses the habits of mathematical reasoning—the subject is one of great importance to me just at present and I should like you to pursue it."[17]

Brunel devoted a lot of his own time to detailed calculations of stresses in bridges, the strength of various forms of construction and similar matters of very practical engineering importance, and he encouraged his assistants to do the same. One who was employed for several years to make calculations and experiments was W. Bell. When Brunel advertised a post for "the superintendence of mechanical constructions" in 1846, Bell applied and Brunel wrote to his referee:

> "Is he industrious and intelligent and secondly is he a willing man or one of those who fancy themselves not sufficiently appreciated—of this latter class I have always a great dread." *(PLB, 6 Jan., 1846)*

Bell appears to have been attached to the work on the South Entrance Lock at Bristol Docks, and was plied with calculations by Brunel on riveting, testing cylinders under stress and other matters. But at the end of 1849 there was little left for him to do in Bristol, although Brunel undertook to retain him on his small staff at a reduced salary to continue some calculations *(PLB, 15 May, 1848)*. Later on he exerted himself in writing references for Bell, who:

> ". . . has been known to me for about 10 years—I have a high respect for his integrity and zeal in the service of his employers—He is a very well informed young man in his profession and particularly also in those branches requiring mathematical knowledge which are too often neglected—he has been engaged on Docks Works as well as Railway construction and if I had an opportunity I should employ him myself." *(PLB, 23 Apr., 1852)*

Eighteen months later there was work to be offered, and Bell was invited to do a job on the South Devon Railway.[18]

How much were Brunel's assistants paid? There seems to have been no standard practice, but the salary varied roughly according to the seniority of the post, the magnitude of the job and the wealth of the company involved. The latter point was important, for although Brunel employed all his own assistants, he expected to reimburse himself for their salaries and expenses from the companies on which he used them. The lowest salaries recorded as being offered to assistants were at the beginning of the series of letter books, when W. G. Owen was offered £150 a year as a "sub-assistant engineer"[19] and Wm Glennie was offered the same terms as Assistant Engineer to be responsible for the Box Tunnel *(PLB, 3 Mar., 1836)*. A more normal salary for an assistant engineer with Brunel was the £300 plus horse allowance accepted by John James on the Oxford and Rugby Railway *(PLB, 31 Jan., 5 Feb., 1846)*. T. E. Marsh had his salary reduced to £300 in 1849: "as the branch to which you are attending is fast diminishing in quantity I cannot ensure you a long continuance at that salary" *(PLB, 10 Dec., 1849)*. H. S. Bush was offered £350 with a prospect of increase to £400 as Resident Engineer on the Cornwall Railway *(PLB, 14 Sept., 1858)*, but on the same day Brunel named the scale £300–450 apologetically to C. Richardson for the Bristol and South Wales Junction Railway, offering the prospect of beautiful countryside and cricket as compensation for the low salary *(PLB, 14*

Sept., 1858). The following day he offered the post as Resident Engineer on a small Welsh branch line to Captain McNair at £225 a year *(PLB, 15 Sept., 1858)*. The work on Plymouth Docks undertaken by Brunel in the early 1850s must have been more remunerative, as he offered his assistant S. Power an increase to £500 a year *(PLB, 24 Apr., 1852)*. At the upper end of the salary scale for assistants, W. A. Purdoe was offered a twelve-month engagement to investigate the possibility of an East Bengal Railway at 1,000 guineas and expenses *(PLB, 3 Nov., 1855)*. Although expenses were normally allowed in addition to the basic salary, Brunel kept a strict eye on this expenditure, as is shown by his reprimand to T. Bratton, his assistant on the Oxford, Worcs. and Wolverhampton Railway, in 1852. He took Bratton to task for charging the *inside* fare on a coach as his travelling expenses when he actually rode *outside*, pointing out that he was on his honour to make a true return and that this device was "*simply not truth*". Finally, he observed that the unfortunate assistant "*should* travel outside—inside is by day—in England—only fit for women and invalids" *(PLB, 7 Jan., 1852)*. As for his own payments, Brunel noted *(PLB, 18 July, 1840)* for comparative purposes that in 1840 Joseph Locke received £300 a year from the Grand Junction Railway with an additional ten guineas a day when employed in company service.

If Brunel appeared anxious to restrict expenditure on salaries to the minimum, this does not seem to have been the case with pupils, where he deliberately charged high fees in order to discourage the numerous applicants who wrote seeking positions in the office of the successful engineer for their sons or protégés. Brunel's attitude towards such pupils is interesting. He did not wish to be bothered with them and made it quite clear that he did not intend to devote any time specifically to them. But he acknowledged a professional duty to have a few such young men in his entourage and laid down strict conditions on which he would agree to tolerate them. Early in the private letter book series he wrote to one parent:

> "I do take pupils—or rather I have been driven to take them—I have no room now and shall have none for 12 months—my terms are 600 guineas . . . P.S. I would not take a pupil without 6 months notice." *(PLB, 3 Oct., 1836)*

Within a few years the range of fees was going up: "the premium would depend entirely upon his (Brunel's) opinion of the qualifications of the young man . . . the premium would vary from £600 to £800 according to the age and term of pupilage" *(PLB, 17 Oct., 1839)*. By 1842 he was charging £800 for a four-year apprenticeship *(PLB,*

28 Feb., 1842), although in the following year he accepted a youth for £700 to cover three years *(PLB, 11 Dec., 1843)*. A year later, however, he answered an enquirer:

> "I diminish the applications by asking what I consider a large premium —£1,000 . . . and I do not profess to give in return anything whatever but the opportunities which my office affords to an industrious intelligent young man." *(PLB, 5 Dec., 1844)*

The fullest statement of Brunel's conception of the role of his pupils occurs in a letter of 1846:

> "I do take pupils but rather as the exception than the rule— that is I do not seek it and rather raise difficulties than encouragement and amongst other difficulties which I create is the premium I charge, I don't know whether it is high compared with what others charge but I know I should think it a great shame if I were a father wanting to put my son with an Engineer—I charge £1,000—that is to say £550 on entering and £150 a year for three years—and one year for nothing making in all four years during which time I profess to take no trouble whatsoever about the youth—he has all the opportunities which my office of course give him—and if he turns out well gets employed in responsible situations which improve him—Do not be induced to expect more for your son but at the same time I cannot deprive myself of the opportunity of saying that it would at all times and therefore in so important a matter as your son's welfare afford me *great pleasure* to forward any views you may have." *(PLB, 17 Apr., 1846)*

The need for pupils to have the qualities of gentlemen was stressed by Bennett, conveying Brunel's terms to an applicant:

> "That the young man shall be of gentlemanly habits as well as of good and gentlemanly connections—and that he shall have had a good education and that either by his special education and his tastes—or by his natural turn and ability he shall have given sufficient reason to suppose that he will succeed in the profession . . ." *(PLB, 4 Nov., 1853)*

It seems likely, given these stringent conditions, that the number of pupils in Brunel's office can rarely have exceeded two or three at any one time, and those whose parents or guardians were able to have paid the fees would have been affluent members of the "gentlemanly" classes.

Brunel's insistence on himself as a "professional man" derived from his pride in his practice of engineering, and he seems always to have taken great pleasure in his intimacy with other engineers. He seems to have attached more importance to his membership of the Institution of Civil Engineers than to his Fellowship of the Royal Society, which he rarely attended; but he frequently took part in meetings of the Civils and it is clear from the patchy correspondence with his fellow engineers that he was often in face-to-face communication with them. Of all these associations, by far the most intriguing is that between Brunel and Robert Stephenson. Born within a few years of each other and dying within a few months of each other the two were exact contemporaries, but their different careers made them strong rivals in the great railway building epoch in which they championed the broad and narrow gauge respectively. Yet in their personal encounters the two men seem to have preserved a strong friendship and even affection for each other. Only hints of these feelings appear in the private letter books, but they are not without significance. In 1844 Brunel wrote arranging to meet Stephenson in South Devon in order to consider Admiralty objections to coastal railways encountered by Brunel in South Devon and by Stephenson in Holyhead. He joked that the objection probably sprang from a fear that the object of coast walling was "to prevent poor shipwrecked mariners from climbing up it" *(PLB, 2 Apr., 1844)*. Replying a few months later to a report that Stephenson had been offended by the hostility of some officers of the GWR, Brunel wrote:

> "I will not conceal from you that the GWR Co. consider the act of the Birmn. Co in going to Parliament for a parallel line of 10 miles as one of the most unprovoked and unmitigated pieces of hostility committed in all this bitter season of warfare—but in the midst of all this warfare we have all sought to avoid anything like personal hostility . . ." *(PLB, 6 Dec., 1844)*

At this time the gauge war was coming to its climax, but a few years later Brunel was still able to write:

> "Excepting on one or two well known points of difference Stephenson and I generally agree perfectly." *(PLB, 18 Feb., 1847)*

And in later years the two men clearly enjoyed being consulted together on engineering matters, as they contrived to go off to various parts of the country to do this, both Manchester and Glasgow waterworks benefiting from their collective advice *(PLB, 1852, 1854)*. They also supported each other in public on a number of important occasions, Brunel backing Stephenson on the controversy about the collapse of Chester Railway Bridge and attending the construction of the Britannia Bridge over the Menai Straits, while Stephenson turned out to support Brunel in the difficult weeks when the *Great Eastern* was being launched.

Towards some other professional engineers, however, Brunel preserved a certain aloofness and even coolness. He referred disparagingly to J. M. Rendel when the Bristol promoters of the Portishead Pier project consulted him without first seeking the advice of Brunel:

> "The person preferred to me was my Surveyor Assistant, a short time back, on the Plymouth Railway, a man unknown . . . a stranger (and) in every respect my acknowledged inferior." *(PLB, 27 Jan., 1840)*

The private letter books also reveal a certain impatience with Sir John Rennie in the course of negotiations which concerned them both *(PLB, 1843, 1845–46, 1851)*, and Brunel expressed robust disagreement with William Cubitt when they served on the Great Exhibition buildings committee *(PLB, 11 Mar., 16 May, 1850)*. But he seems to have accepted Paxton's Crystal Palace design readily, despite having his own ideas for an appropriate exhibition building, and in later years gave Sir Joseph Paxton advice and practical engineering assistance on the construction of the two water towers when the palace had been moved to its new site *(PLB, 1853)*. The private letter books are peppered with correspondence with these and other engineers, such as Vignoles, Locke, Nasmyth and Scott Russell. On the last of these, there is little indication of the struggle which surrounded the construction and launching of the *Great Eastern*, and it seems likely that any written evidence of this was destroyed by Brunel's sons (Rolt, 1957, p. 324). But on a happier note, he corresponded cheerfully *(PLB, 7 Feb., 1853)* about rifles with octagonal barrels, spoke warmly *(PLB, 21 July, 1855)* of W. J. Armstrong, who was later suspected of appropriating such a device (I. Brunel, 1870), and wrote glowing testimonials for his friends such as that which helped Joseph Bazalgette to get the job with the Metropolitan Board of Works which revolutionised the sewage disposal system of London *(PLB, 10 Jan., 1856)*.

Central to this web of friends and professional contacts was membership of the Institution of Civil Engineers. Several communications with Charles Manby, the reforming Secretary of the Civils who established the practice of electing a president each year, are recorded. At the end of 1840 Brunel refused to give Manby some professional information because "it would be a bad precedent and the practice would put a stop to the free and liberal communication now existing in the profession". But he went on to make a concession to Manby:

> "Your reproach upon me for never doing anything for the institution is only too true . . . I will try to do something but you must be easily contented." *(PLB, 5 Dec., 1840)*

Years later he protested vigorously to Manby against a paper to the Institution by Mr Henderson (1853) "On the Speed and Other

Properties of Ocean Steamers", as he took this as a personal attack on his *Great Eastern* project:

"If when Engineers are engaged on new works of magnitude and difficulty which they do not themselves feel yet to be sufficiently matured to bring before the institution and with which the Institution has nothing to do they are to be subjected to the nuisance of being either involved in a wordy war with *all comers*—and the nuisance and inconvenience of having to stand up as a cock shy for anybody who chooses to have a fling, without even the preliminary of paying a penny, or to be subject to the imputation of retreating from discussion—the meetings of the Institution will become a serious inconvenience not to say a nuisance and certainly not a means of encouraging and promoting improvement." *(PLB, 21 Nov., 1853)*

This outburst was followed by another the next day:

"I assure you that I find no fault with the discussion or with anything that I have heard of as said—it was the *printed* invitation to come and see the fun and the promise of a set to between fancy Sam and bloody Bill one of whom at any rate had never wished to fight and has a wife and family dependent on his keeping off the stage—that I complain of—if it is to be a practice. It is not my vanity that makes me believe that everybody understands the 'proposed large steamer' to mean 'Brunel's absurd big ship', but I will bet you any odds that out of any 100 men going into the room of whom you would ask the question, 95 would not dream of any other meaning—and this I say if it is to be a practice will make our meetings nuisances and drive everybody away who has anything new and difficult in hand." *(PLB, 22 Nov., 1853)*

The vehemence expressed here derived from Brunel's deep attachment to the Institution as the collective expression of the profession, and he retained this loyalty to the end. He declined invitations to join other associations which he thought would compete with or injure the Civils[20] and for this reason he observed the foundation of the Institution of Mechanical Engineers with some suspicion. He acknowledged respect for George Stephenson, whom the Mechanicals had been established to honour,[21] but replied coolly to the letter inviting him to join the Mechanicals:

"I beg to request (record?) my thanks to Mr. McConnell and the other members of the Committee of the proposed Institution of Mechanical Engineers for the honour they have done me. Will you oblige me by informing me whether the Institution is proposed to be of a local character or as an Institution for England generally as in the latter case I fear it would tend to create a division in our Institution of Engineers and so far would I think be open to objection." *(PLB, 6 Jan., 1847)*

For Brunel, the profession of Engineer remained a unity, and he did not care for any move to rend what he saw as a seamless fabric.

While Brunel saw himself primarily as a professional Engineer, he mixed easily with men of other professions. He was elected a Fellow of the Royal Society in 1830 (I. Brunel, 1870, p. 516) and became a member of other scientific societies: he was invited to join the Geological Society; he was invited to write some cyclopaedia articles for the Society for the Diffusion of Useful Knowledge; he communicated with Faraday *(PLB, 18 June, 1847)*; he made arrangements for the mathematician and astronomer Airy to take measurements at the site of the Chepstow Bridge[22]; he was a friend of another distinguished mathematician, Charles Babbage, whose son acted as his representative and chief assistant on most of his Italian projects; and he took a continuing interest in the practical application of the electric telegraph, coal-tar derivatives and a wide range of other inventions. Indeed, Brunel was constantly pestered by inventors to comment on their ideas, but he usually refused to do so unless the invention was already publicly established as he had a life-long distaste for patent legislation and did not wish to get involved in litigation with inventors who might claim that he had stolen their ideas. For when he recognised a good idea, like the screw propeller or the principle of atmospheric propulsion, he usually found an opportunity to apply it.

Reflecting on the place of the engineer in society, Brunel had emphatic views about the crucial subject of technical education:

> "I must strongly caution you against studying *practical* mechanics among French authors—take them for abstract science and study their statics dynamics geometry etc etc to your heart's content but never even read any of their works on mechanics any more than you would search their modern authors for religious principles. A few hours spent in a blacksmiths and wheelwrights' shop will teach you more practical mechanics —read *English* books for practice—There is little enough to be learnt in them but you will not have to unlearn that little." *(PLB, 2 Dec., 1848)*

One could hardly hope for a better epitome of the mid-century British view of the best form of engineering education.

Brunel was against state intervention. When the government set up a commission to investigate the standards of the engineering profession after the collapse of the Chester Railway Bridge, Brunel referred to it rudely as the "Commission for stopping further improvements in bridge building and all other applications of iron" *(PLB, 16 Nov., 1847)* and he addressed William Cubitt as "one of the despots themselves on the subject of this iron commission" *(PLB, 14 Oct., 1847)*. He strenuously resisted any effort to lay down conditions for the engineering profession, submitting a memorandum to this effect to the commission:

> "Nothing, I believe, has tended more to distinguish advantageously the profession of Engineering in England and America—nothing has conduced more to the great advance made in our profession and to our pre-eminence in the real practical application of the science than the absence of all 'regles de l'art'—a term which I fear is now going to be translated into English by the words 'conditions to be observed'. No man, however bold, or however high he may stand in his profession, can resist the benumbing effect of *rules* laid down by authority."

He goes on to argue that improvements in technique may lead to great improvements in the quality of large iron castings, if this is not prevented by restrictive legislation. Any such restriction "is contrary to all sound philosophy, and will be productive of great mischief in tending to check and to control the extent and direction of all improvements as preventing that rapid advance in useful application of science to mechanics which has resulted from the free exercise of engineering skill in this country—subjected as it ever is under the present system to the severe and unerring control and test of competing skill and public opinion; and, devoted as I am to my profession, I see with fear and regret that this tendency to legislate and to rule, which is the 'fashion' of the day, is flowing in our direction" *(PLB, 13 Mar., 1848)*.

With this robust and articulate statement by Brunel when he was at the height of his powers we may bring this chapter to a conclusion. Brunel was not saying less than the truth when he described himself as being devoted to his profession. He applied almost all his incredible energy throughout his adult life to the pursuit of his engineering commitments. He refused to be distracted by the temptations of a political or a social life, and found little time for recreation, holidays or spiritual activities. By maintaining rigorous professional standards for himself and his assistants he was able to assert a remarkable degree of independence in relation to the many boards of directors which he served, and thus helped to establish the role of the engineer as a professional consultant. He turned down most invitations to office in societies and frequented none except the Institution of Civil Engineers, and when his turn came to be President of this following Stephenson's term of office he was obliged to postpone the opportunity on account of his failing health (I. Brunel, 1870, p. 516). He looked askance at state honours in England, where "merits can be so much better and more surely marked by public opinion" *(PLB, 9 Feb., 1856)*. The only honour he sought was the recognition of his works. The best valediction to him was thus that inscribed in 1859 on the portal of the Royal Albert Bridge over the Tamar, "I. K. BRUNEL: ENGINEER".

II

Tunnels

SIR HAROLD HARDING

Tunnels played a vital role in the life and work of Brunel. They fall into two contrasting periods. The first was that of his deep involvement in his father's famous Thames Tunnel, while the second covered his own railway tunnels. Sir Marc Brunel was a pioneer in developing shields for tunnelling. He took out a patent in 1818 for a circular cast iron shield propelled by hydraulic jacks with a brick or "preferably" cast iron lining, and in 1824 his design for the Thames Tunnel shield was "brought before the Public" at a meeting of the Institution of Civil Engineers. In this same year Brunel entered his father's office at the age of seventeen and became a party to all his father's thinking.

The Company to carry out the Thames Tunnel was formed in 1824 and work was started the next year on the shaft at Rotherhithe. At the age of eighteen Isambard began, with two other assistant engineers, at the very start of the work. In 1826 Armstrong, the Resident Engineer, broke down and resigned, so Isambard became Resident Engineer in his place. To prevent him working himself to death he was given three assistants, the most important being Beamish, a 28 year old ex-Guards officer. Isambard was involved day and night in every detail of the work, and as it was all carried out by direct labour he learned all the duties and responsibilities which fall to a modern Contractor's Agent. He learned how to handle labour and the importance of ensuring the supply of materials as well as the programming of the day-to-day work and the paying of wages. The Thames Tunnel was the first of its kind, so father and son were learning as the work progressed, although the foresight shown by the father was quite outstanding. Experience is measurable by intensity and not by time and young engineers who start on tunnels spend most of their time at the working face, an experience that proved of unique value to Isambard's later work. One of the first lessons to be learned was the need for accurate site investigation which Isambard insisted on in his later tunnel work. He was far in advance of his successors as 100 years were to pass before this became a scientific study. Geology was then still in its infancy: William Smith had issued his first geological map only in 1815.

H. Law (1845) records that Sir Marc employed two engineers, Joliffe and Banks, to make soundings across the river at the chosen site between Rotherhithe and Wapping. They worked from a small boat and tested the material in the bed of the river by plunging a 2 inch diameter pipe into the river bed. Not much reliable information could have been gained by this method in a swift-flowing tidal river carrying much shipping. Law, who worked on the tunnel, quotes from the Directors' first report:

> "They now have the satisfaction to inform you that the result of thirty nine borings made in two parallel lines across the river, has fully confirmed the expectations previously formed, there having been found upon each a stratum of strong blue clay of sufficient depth to ensure the safety of the intended tunnel."

This may have implied that there was clay cover over the top of the tunnel as the soundings could have hardly revealed the full depth of the face. In 1807 Vaizey and then Trevithick had driven an exploratory timbered heading 5 feet high and 3 feet wide for 800 feet under the Thames at the same site until water caused the heading to flood and collapse. Sir Marc would have known of the strata which they met, which were principally the upper layers of the Woolwich and Reading Beds, in which his tunnel was eventually driven. In the same beds the Rotherhithe Tunnel was subsequently driven using compressed air and, many years later, the Central Line tube railway extension under Stratford Marshes. Fig. 2 shows a section of the tunnel shaft taken from Law's memoir and the strata which he recorded. Sir Marc had been warned of "a quicksand of a very dangerous character with copious springs". This is the Thanet Sand which has made many difficulties for workers in that area; it prevented him from sinking his sump to the depth which he desired and caused him to start his tunnel 5 feet higher than he had intended. But there is little doubt that Sir Marc still had confidence that his shield could cope with these conditions and he proved right, as although the tunnel was flooded three times it was never "lost", nor did it collapse, and the shield remained intact.

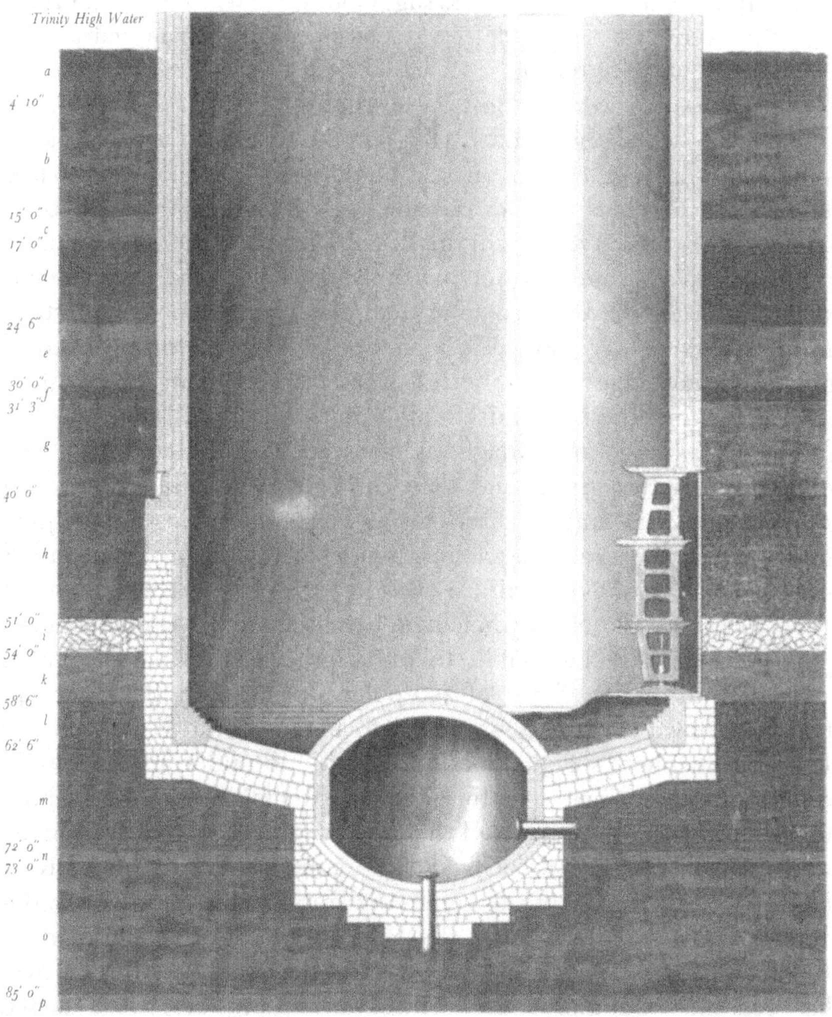

Key. a: Loam and made ground. b: Gravel, firm and free from water. c: Loose sand full of water. d: Very coarse gravel with large stones; "water so considerable over powered steam engine". e: "Quicksand" in flux equally abundant. f: Very base gravel; little water. g: Very stiff blue clay (London Clay). h: Clay more mottled in character (Woolwich and Reading beds); portion of silt; great number of shells. i: Chalk and clay combined; chisels to break out. k: Very firm green sand and gravel. l: Greener gravel and sand with chalk; red ochrous earth; a little water. m: Sandy loam quite free of water. n: Quicksand "copious in flux". o: Very green gravel; very little water. p: Light slate-coloured quicksand of very dangerous character; springs copious.

Fig. 2. Thames Tunnel: sectional elevation of the Rotherhithe shaft

Sir Marc was acutely aware of the importance of the nature of the ground. In 1828 after his tunnel had been brought to a stop he received many letters with useless suggestions. He wrote in his journal:

> "In every case they made the ground to suit their plan and not the plan to suit the ground."

These wise words were not lost on his son.

In 1825 work began (Law, 1845) on the Rotherhithe shaft, based on well-sinking methods but on a far larger scale. The shaft was 50 feet in diameter to allow room for the shield. On a circular iron kerb brick walls 3 feet thick were built in "best Roman cement". Portland cement derived from this but had only been patented the previous year. The walls were raised to 40 feet and strengthened by vertical iron rods which were bolted to a timber kerb at the top to secure it to the cutting edge kerb. The full story of the sinking of the shaft reveals the fine foresight and attention to detail of the Engineer. The shaft sank under its own weight as men excavated below the cutting edge through dry gravel and then through the coarse gravels and sands where "the influx of water was so considerable that it quite overpowered the steam engine and rendered the aid of hand pumps necessary". Luckily very stiff clay was met with from 31 feet to 40 feet down. The outside of the shaft had been rendered with 2 inches of "half cement and half sand" to reduce the friction which overall amounted to the low figure of 350 pounds per square foot. From 40 to 50 feet down clay of a more mottled character (the Woolwich and Reading Shepherd's Plaid) made it possible to underpin the cutting edge and, with various adventures due to water following down behind the brickwork and stoppages of the steam engine, the shaft was finally underpinned to a depth of 65 feet. Young Isambard was actively engaged in every step of this unique operation. The beds below this clay exhibited all their usual characteristics as some contain much water and some do not. At the bottom of the shaft Sir Marc formed his "cess pit" or sump in careful brickwork in and below the domed masonry invert, to a depth of 80 feet. This was not as deep as he had intended due to the water-bearing ground. Below the clay the bed of limestone which is common to these beds was encountered and it continued through both this and other tunnels. The excavation of the cess pit was a battle in itself with the upsurge of groundwater, using wells sunk as sumps, primitive sheet piling and other methods. He had no compressed air, groundwater lowering techniques or chemical injections to help him.

When the shaft had been finally completed a square timber

frame was built up from the base, and the top of the shaft was covered by a substantial deck which also had to support the 30 horsepower steam engine. This engine had to work four plunger force pumps, one at each corner of the frame, as well as an endless bucket chain for bringing the spoil to the surface, which could also deal with water. The four pumps together could handle 500 gallons per minute and a weir with a 12 inch square notch was built into the flume which led to the river in order to measure the quantity of water pumped. This was one more example of Sir Marc's attention to detail, not lost on his son. This was a valuable rehearsal for Isambard's future bridge foundations, as at Hungerford and Saltash. Rather primitive steam engines were used and simple pumps, not always the ones which his father had ordered. But much was learned of the possibilities and limitations of mechanical means and the problem of pumping from water-bearing beds.

In October 1825 the parts of the shield began to arrive at the site and the shield was erected inside the shaft. The shield was the vital key to the whole venture. Sir Marc was never mean in his conceptions and he wisely considered that the whole project would be of little value unless the capacity for traffic was as large as possible. So he designed twin carriageways, each 14 feet wide and 15 feet 3 inches high. He had given his usual deep study to existing methods and experience of others in tunnelling in soft ground. His work at Chatham brought him into contact with the Army, whose Sappers and Miners were also trained in the accepted methods used by Vaizey and Trevithick. In his book on fortifications Professor Muller (1770) of Woolwich wrote:

> "The Gallery being carried on some distance, carpenters follow the miners to prop the earth above, if it is not stiff clay or loam, to prevent its falling in. This is done by placing a piece of wood a-cross the Gallery overhead, and putting a prop or upright at each end, fastened into the ground, then the earth is removed above this place, so as to slip a piece of deal board over it, which being afterwards supported in the same manner at the other end will sustain the earth."

His successor, Isaac Landmann (1815), writing on mines shows a timbered heading almost identical with the headings in loose sands and gravels which were driven to build the walls of Piccadilly Circus station in 1928.

The principle is to expose the ground as little as possible. Overlapping timber "piles" of 1½ inch boards placed over the head trees and behind the side trees are tapped ahead while the miner scratches away the loose ground. The face is held by horizontal boards which are removed and advanced, one at a time, as the face is worked down. Such a heading contains a miner and a miner's labourer, with labourers backing up behind. A good miner never takes his eyes off the face and calls to his mate for "wedges, long litter, club hammer" as a surgeon calls for "scalpel, forceps, wadding".

Sir Marc's shield (Fig. 3) was designed for operation on the same lines. The first shield was 35 feet wide and 20 feet 6 inches high. It consisted of twelve vertical cast iron frames, side by side, each with three compartments, one above the other. On the outer frames and over the top of all the others, carefully designed cast iron units were placed which worked on rollers and fulfilled the functions of the normal timbers: they were worked forward as excavation continued in order to protect the roof and sides while the whole face of the tunnel was supported by timbering. This took the form of horizontal boards, 6 inches high, 3 inches thick and 2 feet 10 inches long. Each frame thus had its own panel of face boards, which were held against the ground by screw jacks at each end. Each board could be advanced one at a time from the top downwards. Each frame (Fig. 3) could be advanced separately by means of heavy screw jacks acting against the brick tunnel lining which closely followed the shield. Thus Sir Marc's shield consisted of the equivalent of 36 timbered headings being driven almost simultaneously. This justified the rectangular shape, which is almost unique. The original plan was to advance alternate frames 4½ inches so that there would never be a gap between the ends of the breast boards. Each board was removed, the ground dug out and the board replaced and strutted off the adjacent frames so that when the face had been "taken down" the frame was free to be pushed ahead and the face boards strutted to it once more. As clay had been expected to continue overhead, the tail plates to overlap the finished brickwork were omitted in the first shield: a step later regretted.

Until the Second World War brickwork was the chosen lining for railway tunnels and for many sewers, until it was killed by economics. Those who have never worked with it would not realise its many virtues and flexibility. The Thames Tunnel was lined with bricks laid without breaking joints as 9 inch stretchers, the arch being six rings thick. When smaller advances were needed the bricks were laid as headers in three rings. This was necessary to produce a flat face for the shield jacks to press against. The spandrels were filled with brickwork in the normal way and the arches between the two

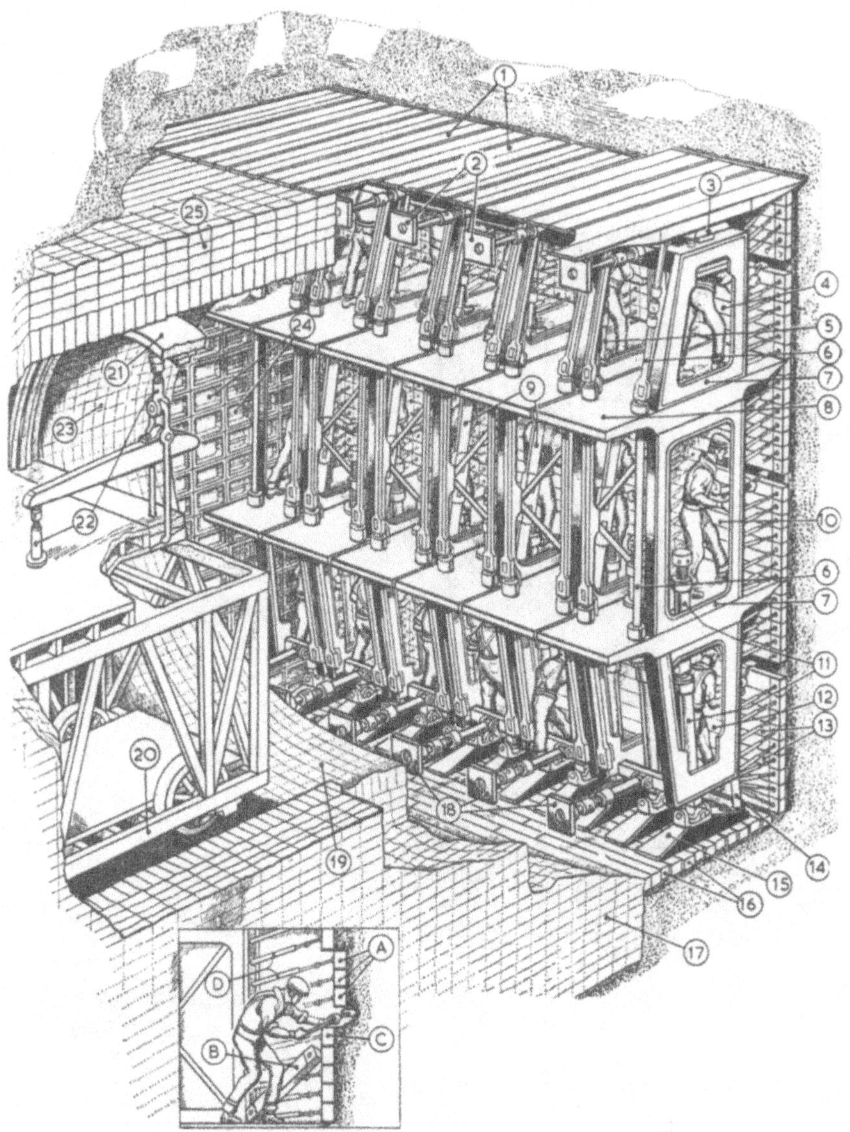

Key. *1: Top staves. 2: Top abutting screws. 3: Head. 4: Top box of frame no. 6. 5: Tail jack. 6: Wrought iron reinforcing member. 7: Cast iron side frame member. 8: Upper floor plate of frame no. 6. 9: Sling. 10: Middle box of frame no. 6. 11: Leg. 12: Bottom box of frame no. 6. 13: Poling boards. 14: Jack forcing down floor boards. 15: Shoe. 16: Floor boards on which brick roadways rest. 17: Brickwork of dividing wall. 18: Bottom abutting screws. 19: Brick roadway. 20: Travelling stage. 21: Roof centring. 22: Jacks for adjusting roof centring. 23: Western sidewall. 24: Side staves. 25: Roof brickwork.*

A: Poling board moved forward. B: Poling board removed so that miner can excavate. C: Poling board that has not been moved forward. D: Poling screws.

Fig. 3. Thames Tunnel: the shield and face in the western archway

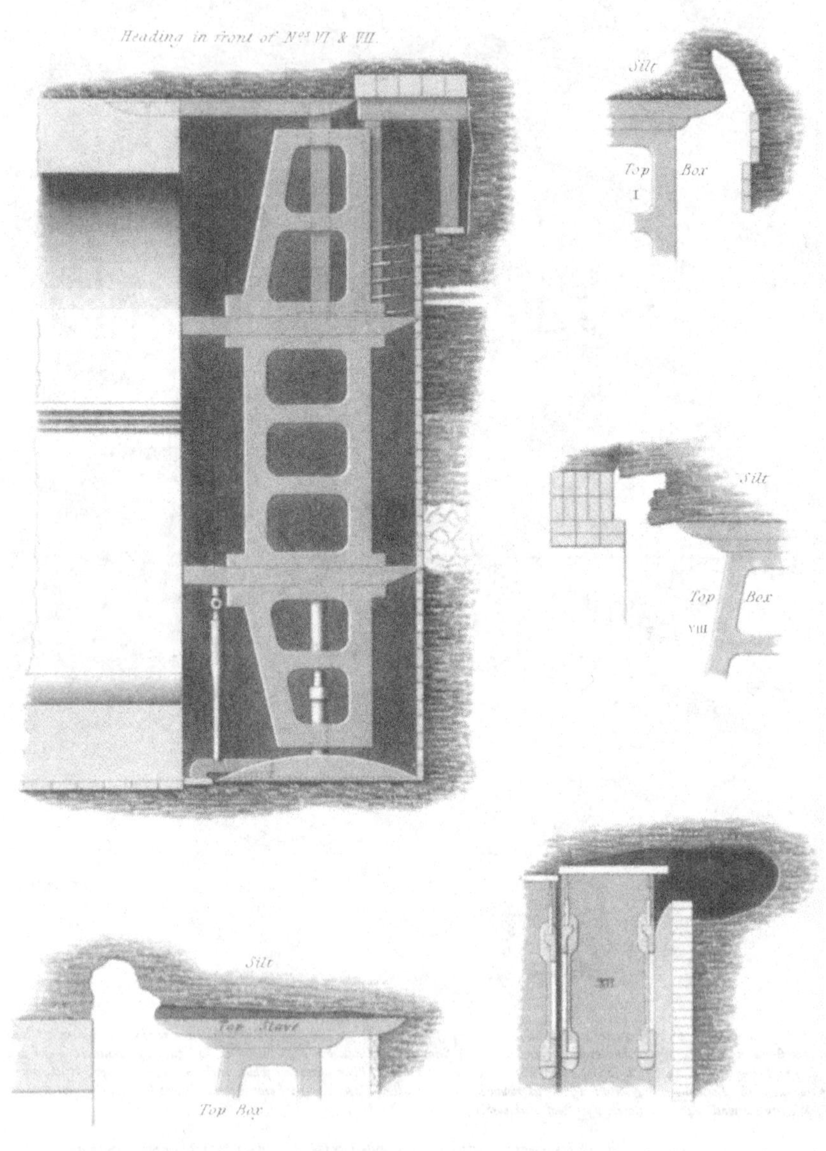

Fig. 4. Thames Tunnel: details relating to the progress of the works

carriage-ways, which are such an architectural feature, were cut out after the shield was well ahead so as not to weaken the centre wall on which the jacks had to press. The shield towed a timber stage for the bricklayers. This towing of a stage behind a shield has only recently been revived! There is little in modern practice which was not foreseen by Sir Marc. The brick lining arches were turned on an iron semi-circular rib carried on a long cantilever arm, one end of which rested on the floor of an upper compartment and the other on a screw jack on the stage.

The "burying" of a shield at the start of a drive is always slow, even with experienced miners. Sir Marc reported to his Directors:

> "I invariably expressed that I considered the getting out of the shaft and securing the work quite into the undisturbed ground as the most difficult point to be overcome, considering the men, by whose agency this was to be effected, were totally unacquainted with the instruments that were to be used, and that too under the most disadvantageous application that can be imagined. No provision could be devised to qualify at once the roughest quality of men, such as miners, bricklayers, and labourers for the management of a machine of such complication in its parts, and of such nicety in its movements as the shield is." (Law, 1845)

The full detailed story of the progress and adventures of the Thames Tunnel and its protagonists has been re-written by many hands (e.g. Beamish, 1862; Clements, 1970) but here it is simpler to consider the progress in terms of Law's report (1845) on weekly progress (Fig. 4). Isambard Brunel was vitally concerned in the first 605 feet of drive, or half the total distance, until the financial collapse and the building-in of the shield at the end of July 1828. Water was encountered early on while attempts were made to train unmechanical and illiterate men to use the ingenious but complex details of the shield. By 16th February, 1826, the shield had become "deranged", but by March the miners were considered to have been trained and, until the end of May, the progress averaged 7 feet per week. In general two shifts of 8 hours were worked, but the number of men and their disposition at this time is not clear. Some 36 men appear to have been involved at the face but there would only be one miner excavating in a compartment with his miner's labourer or mate. If the ground was loose only one board at a time would be worked—down from the top to bottom in a frame. If the ground was firm then work might proceed at all three levels in a frame at the same time, although only every other frame would be worked at one time. But there was plenty of work other than excavation to be done as the staves at top and sides would have to be advanced to keep pace and there was much else needing attention.

Throughout the work Sir Marc suffered from the ignorant interference of some of his Directors, especially of the Chairman. One decision which he resented all through the work was the refusal to allow him to make a drainage culvert to conduct groundwater away from the face as the tunnel was being driven downhill. He described this as a convenience never lost sight of in mining. The provision of a sub-drain below brick-lined sewer tunnels was later to become standard practice by the London County Council until this type of construction was abandoned in recent years. Then again the Directors forced him to double the advance of each frame. This opened up gaps at the ends of the face boards and allowed ground to enter which was a matter which the design had been intended to prevent.

In June 1826 the shield had to be stopped for a week owing to the men's failure to follow instructions, "while the frames were pulled together". This simple statement does not adequately describe a remarkable feat. After a timbered chamber had been dug for 2 feet 3 inches to the north to accommodate it the whole massive contraption was pulled together by chains and moved bodily sideways while still sustaining the earth pressure. Cast iron is a brittle material but has the virtue of being quickly replaceable as long as the patterns are kept available, so it was possible for some cracked legs to be replaced at the same time. Such replacements occurred frequently during the work.

Then on 7th July Sir Marc had to give way once more to the pressures of his Directors, against his better judgement, and allow piece-work. The bricklayers egged on the miners who controlled their progress, so the miners worked recklessly. For a brief spell progress improved but after a month, as a consequence, the shield again became completely deranged. This allowed gaps to appear between the frames as well as in the face and so caused the shield to lose some of its protective value. Piece-work was abandoned in November 1826 and progress improved in spite of worsening ground conditions. So with many ups and downs the tunnel went ahead at an average rate of 9 feet per week while a rate of 14 feet was reached for several weeks.[23] The water problem was constant, much of it coming from the groundwater in the lower beds of silt and sand. Owing to the lack of a sub-drain, water accumulated at the tunnel face and had to be removed by hand pumps. In March 1827 the report mentions twelve men working three pumps.

Just before the first inroad of the river occurred there were 40 men pumping from ten pumps and they had to work from the level of the decking of the top stage so that the drop of 14 feet could carry the water back down the sloping timber chutes to the shaft sump. In spite of the inflow of water the greatest quantity pumped until the first inroad was measured only as 500 gallons per minute. Today this is easily dealt with by a centrifugal pump of 6 inch suction. This shows the difficulties which had to be overcome when no such pumps existed. In January 1827 the Directors wanted to turn the work over to contractors. Brunel reported that, "We have been visited by almost all the tunnel makers in the country", and strongly urged the impossibility of a sensible contract being obtained. As the railway age was only beginning it is a matter of speculation how many tunnel makers existed.

Isambard was now officially appointed as Resident Engineer at the age of 20, although he had been largely responsible for the day-to-day work in the face and for urging on a few older men. He had already earned a reputation for his energy, his ability, his unquenchable optimism, his heroism in rescuing men on several occasions and his survival after serious accidents. Throughout his work on the tunnel and later, listening to his father's further experiences, Isambard became acutely aware of the stupidity and obstruction exhibited by the Directors, especially the Chairman, and the patience with which his father overcame these hurdles. When we read of the shield becoming deranged it is a wonder that the Brunels and their devoted assistants did not suffer the same fate. Isambard was to meet with plenty of such obstruction and human weaknesses in the years ahead and benefited by example and experience. Criticisms by amateur and armchair "engineers" still continue!

Progress was steadily maintained until the end of April 1827 in spite of drunkenness among the men, strikes and worsening of the ground. On 22nd April a diving bell was borrowed from the West India Dock Company and both the Brunels descended to examine the effect of the tunnelling on the river bed. What followed then has been described in many popular publications. The water burst in, the tunnel was flooded and Isambard carried out rescues with gallantry. The tunnel was then 549 feet long. From 15th May to 28th September the time was spent in trying to close the cavity in the river bed, and when this was successful work began on restoring the shield and the conditions below ground. By 28th July the tunnel was again lit by gas, which betokened success, and new castings were built into the lower boxes of the shield. The drive was renewed on 28th September.

With the tunnel flooded, Isambard had a period of work in the open air which provided completely different experiences. What he learned was the difficulty of organising work in a fast flowing tidal river which carried much shipping under sail. The Thames Navigation Committee had to be appeased, arrangements made to hire a crane barge to handle the diving bell and moorings to hold it in position. Fortunately diving bells had been developed by John Smeaton for use in harbour work, although the compressed air had to be pumped to the bell through leaky leather tubes. Careful surveying was needed to locate the bell over the tunnel face (Fig. 5) and the descents in the diving bell revealed cavities over the tunnel shield which could itself be seen exposed to the river water.

Sir Marc, as well as his son, made many descents and plans were made to fill the cavity. The statement that bags filled with clay were dumped is an over-simplification; clay had to be located, dug and transported and then filled by hand into bags. Bags had to be found and collected. On one occasion 21,000 cubic feet of clay in bags, skewered with stout hazel twigs, were thrown in. This meant as many bags, the hiring and loading of barges, and their towing to the place, mooring and discharging before steam tugs became commonplace. Several attempts were made by this method, and then by the use of a timber raft sunk by being weighed down with bags and gravel, which tilted and failed. Next a tarpaulin, 80 feet by 80 feet, was spread and lowered over the cavity. To support and stretch the tarpaulin in the tidal current must have involved several craft, especially as the edges were bound with 9 tons of chain.

So, largely by the efforts of Isambard, the cavity in the river bed was choked and the tunnel pumped out. At the end of 1826 he had written (Rolt, 1957), "After all I may be said to have almost built the Tunnel", and this is a fair assessment. After his adventures with floating craft on the river he wrote:

> "I cannot help observing, for my future guidance, that being alone, and giving few but clear orders, and those always to the men who were to execute them, I succeeded in an operation not altogether mean, and which a very trifling want of precaution or order might have caused to be a total failure." (I. Brunel, 1870)

The "future guidance" of alone giving the orders was followed by him when he floated out the Chepstow truss, and the Saltash caisson and truss. We are told little about Hungerford Bridge, but the piers of

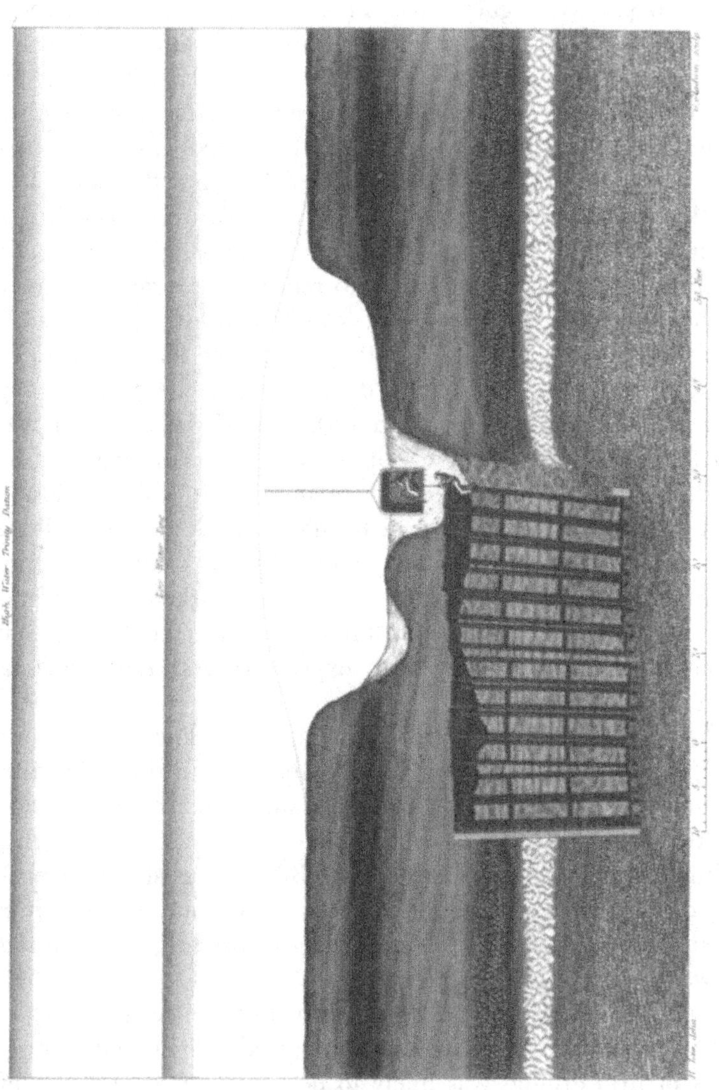

Fig. 5. Thames Tunnel: inspection of the river bed

brickwork formed inside cofferdams of timber sheet piling called once more for work in the tidal Thames.

Work was resumed, but with slower progress, and Isambard's active mind led him, on 9th November, 1827, to give a dinner inside the tunnel to 40 invited guests and, in the neighbouring arch, 120 of his workmen. But when a total length of 605 feet had been driven on 11th January a second break-in of water occurred. The diving bell was not immediately available and the state of the river bed was not known before this disaster took place. Isambard was in the shield when the water burst through the timbered face and the side staves. The gas lights were blown out and all at the face were struggling in the water. This time he was pinned under the staging and badly injured his knee and so was not able to repeat his swimming feats of the previous occasion; six men were drowned and he only escaped by being washed to the foot of the shaft. His absence for several weeks was deeply felt by his father but, in due course, all the previous steps were repeated to shut out the river and the tunnel was again emptied of water. But the coffers of the Company were also emptied, so there was no alternative but to secure the face and frames of the shield and to wall it in with solid brickwork in Roman cement. So ended Isambard's part in this world-famous work. Today Outward Bound courses are organised to help youths to "develop character". Isambard certainly had a four-year course of intensive character development and emerged at the age of 22 with experience and reputation far beyond men of double his age.

So by August 1828 Isambard Brunel had to seek work elsewhere. Rolt (1957) has called this period the "years of frustration", but they were very active. Then, in 1833 Brunel was appointed Engineer to the future Great Western Railway. For two years he was fully stretched in battling through the Parliamentary Bills while simultaneously carrying out the extensive and detailed surveys for the route from London to Bristol and pin-pointing the lengths where he decided that tunnels would be needed.

Brunel had to adapt his hard-won experience to different methods. His railway tunnels would not be under rivers but through hills; rock or firm ground was likely to be encountered and not soft material, so it is doubtful if he ever contemplated using a shield. During the canal age many tunnels had been driven—although they were small in size—such as the Sapperton Tunnel on the Thames and Severn Canal which was opened in 1789 and is over 2 miles long. The method of tunnelling which was then perfected and later used on innumerable railway tunnels persisted well into this century and is known as the English method, or by miners as "tunnelling on bars".

Brunel's notes and sketches on tunnels are sparse and spasmodic. No doubt he had little time for making notes as fully as he did on the Thames Tunnel with only one face at work. In his documents which are at Bristol University a number of small references are made to tunnels and these are less informative to the reader than to Brunel. There is one page *(Bristol Railway)* headed "Tunnels—Bristol" with several columns, first giving the name of the rock and then area by length to give cubic yards, and then these are monied out at simple unit rates. He listed five tunnels. Tunnels 1 and 2 in Red Sandstone, 3 and 4 in Pennant and 5 in Lias. He arrives at a total cost of "£31,191–8–2½" by means which are not too clear except to himself. His estimating seems simpler than in modern methods but is probably rough checking. For example:

> "*Bristol Railway.* Suppose through rock. tunnel 35 ft wide 35 ft high then o = 40.74 Sq. yd.
>
> "Now in good Bath Stone this work would certainly be executed at 3/- per yard excavated and 2/6 per sq. yard dressing. ? too high. Say 4/- incl. dressing × 110^{y} = £20 per yard = £35,000 per mile."

There are several half-finished sketches of "Tunnel Fronts", headed "Bathwick No. 3, No. 3B", "Tunnel mouth at Saltford", and "Twerton Tunnel", dated 28th December, 1836. The Twerton portal is a fine ink drawing *(SB, 1836)* in some detail and shows two Gothic arches, one over each track inside a larger arch with a rose window in the space above the twin arches. On either side are stepped buttresses and the whole effect is distinctly ecclesiastical. Mixed in with these notes are some on tunnelling methods used near Brussels and a sketch of the Belgian method of timbering, using radial supports. In addition to the notebooks, a number of Brunel's original letters are in a collection covering this period. They are written in a rapid sloping script, using the old-fashioned long s in the middle of words. He did not allow punctuation to interrupt his flow of instant but clear thought. His contract drawings are preserved at Paddington, hand coloured and with fine steel engraved draughtsmanship, giving detailed drawings of these and other portals.

He wrote of contracts Nos 1, 2, and 3B in his report to the Directors *(PLB, 13 June, 1836)*. According to the *History of the Great Western Railway* (MacDermott, 1927), starting from Bristol there are the following tunnels: No. 1—1,236 yards long, No. 2—154 yards, No. 3 (Brislington)—1,017 yards and then two short tunnels leading to the Twerton tunnel of 264 yards. Brunel wrote in his report:

> "No. 1B which includes the greatest portion of the tunnelling and generally the heaviest task which occurs between Bristol and Bath . . . has been contracted for by Mr. Ranger . . . a great quantity of the best material derived from the excavation has been carried through the tunnel to the works at the other extremities of the line, and consequently the forming of a heading or driftway through the line of these tunnels as the first and most important operation—and the period for opening these headings was limited on the contract under heavy penalty . . . for securing proper expedition in Tunnel No. 1 (the first from Bristol) two temporary shafts have been sunk so that the headings can be carried on at 6 places at once and from each end and two from each shaft—In Tunnel No. 3 which is the longest between Bristol and Bath being upwards of half a mile, three permanent shafts have been sunk and two temporary ones so that the headings can be worked at 12 different faces; at 8 of these they are active in operation and the remaining 4 will be very shortly, the two shafts from which they will be worked being within a few feet of the required depth, at the headings the work has been carried out on day and night but I do not expect that they will be completed within the time fixed by the Contract which for the tunnels No. 1 and 2 expires on the 21st of next month, time has been lost in this & in other parts of the work by the injudicious arrangements of the Contractor's foreman, at my request he has been dismissed and the work has since proceeded more expeditiously and with greater advantage to the Contractor himself as well as the Company."

This report seems to be in Brunel's more careful hand but composed with his usual breathless haste and little punctuation. The report presupposes a large number of tunnel gangs working at one time when railway tunnels were still in their infancy. The gangs of miners at work at one time would be in only half the number of faces so as to allow for the entry of the bricklayers. It may be that the haste on these lengths was to free gangs of miners for the Box Tunnel, which was delayed by shaft sinking work, and this may account for the opening of so many faces. He expected this work to be completed early in 1838.

The Box Tunnel was the key to his critical path for the opening of the line from London to Bristol and at the time was thought remarkable for its size and length. Unlike the Thames Tunnel all the work was let out to contractors of varying competence. Some of the details of this work apply to other of Brunel's tunnels and some published accounts differ, but the original contract and contract drawings are available for study at British Railways at Paddington. From the start Brunel was plagued by critics whose only qualifications were a classical education and a desire to become amateur engineers in those days of novel and rapid expansion; but Brunel defeated them all. Pudney (1974) and other writers have described the panic which seized the critics at the thought of passing through "that monstrous and extraordinary, most dangerous and impracticable tunnel at Box". Yet since 1830 trains had run through tunnels on the Manchester–Liverpool line. The Kilsby tunnel was being driven by the Stephensons through more difficult ground with much less excitement.

The work started in 1836 as shown by Brunel's letter (private letter book, *PLB, 3 Mar., 1836*) to William Glennie who had applied for the post of Resident Engineer. In a long letter he said that Glennie must prove himself step by step, and added:

> "My responsibility is too great to allow of my retaining for one moment from any feeling of personal regard, the services of any one who may appear to me to be inefficient from any cause whatever and consequently it is an understood thing that all under me are subject to immediate dismissal at my pleasure you will perceive that I state all these conditions in strong and perhaps harsh language and that in substance they are exactly what I stated to you."

Brunel's idea at the age of 30 was to "start as you wish to go on". Glennie survived the testing time and proved an efficient Resident Engineer throughout the work.

In June Brunel wrote a five-page report *(PLB, 13 June, 1836)* to the Directors of the Great Western Railway from his house in Duke Street, Westminster. Four pages dealt with progress from the Bristol end, including several tunnels on the way to Bath. Then on page 5 he wrote:

> "Beyond Bath the Box Tunnel is the most important work . . . will determine the completion of the whole line—active steps are consequently being taken to put this part of the work in a train for proceeding. . . .
>
> "Five temporary shafts have been sunk on the line of the tunnel to various depths varying from 40 feet to 90 feet to determine the position of the strata of the Oolite through which all of them have been carried —a sixth has been found necessary at the west end—before I can determine with sufficient certainty the exact position of the clay and Fullers Earth which lies under the Oolite and the proportionate length of the tunnel (which it) will pass through must govern the relative distance of the permanent shafts—this remaining shaft will be worked day and night and as soon as the required information is obtained which I hope will be in a fortnight, we shall be able to prepare and to let the contracts

of these permanent shafts which I propose to do separately from the Tunnel in order that the materials through which each portion of the latter is to be carried may be ascertained and worked by the parties most accustomed to the particular description of material and not contracted for blindly on a mere speculation.

"I am genln your most obt. st.

I. K. Brunel."

So once more this admirable engineer is years ahead of his time by insisting on a sensible site investigation so that he could design the work on the lines of his father's famous dictum "to make the Plan to suit the Ground". His merit over this has never been properly applauded and should be more widely known. Then on 10th October Brunel wrote to W. Palmer Knight Esq.:

"At the request of Messrs Orton & Payton I have to inform you that they have obtained a contract with the Great Western Railway Company for sinking seven shafts for the Box Tunnel near Bath, which contract, I should think would, with the Works connected with it, amount altogether to about £20,000." *(PLB, 10 Oct., 1836)*

By the following April troubles had arisen and Brunel wrote to W. Tothill, Esq., Great Western Railway office, at some length. He asked for the Committee's instructions on his proposal to pay off Payton and keep on Orton, with assistance to prevent the public breaking of their contract. Brunel was quick to forestall the many critics and his letter went on, ". . . it will be a liberal act and whether the advantage to be gained in preventing unpleasant though inaccurate reports is sufficient to justify such liberality the Directors alone must judge and determine" *(PLB, 12 Apr., 1937)*.

The shafts were completed in time for letting contracts for the tunnelling in 1837 and 1838. Fig. 6 shows a longitudinal section of the Box Tunnel derived from the section on the contract drawings at Paddington for contract No. 4, while the geology is taken from a section supplied by the Institute of Geological Sciences. The position of the seven working shafts was shown on the original documents while the extra shafts were found plotted on a progress plan made on a print from the contract drawing but with no dates. This latter plan shows that tunnelling was being carried out in both directions from every shaft and also at the portals. The main shafts are shown on the contract drawings as 25 feet internal diameter with differing thicknesses of lining for different strata but with no written explanation on the drawings. One coloured drawing shows how the shaft was to be turned by expert brickwork into a smooth transition from the

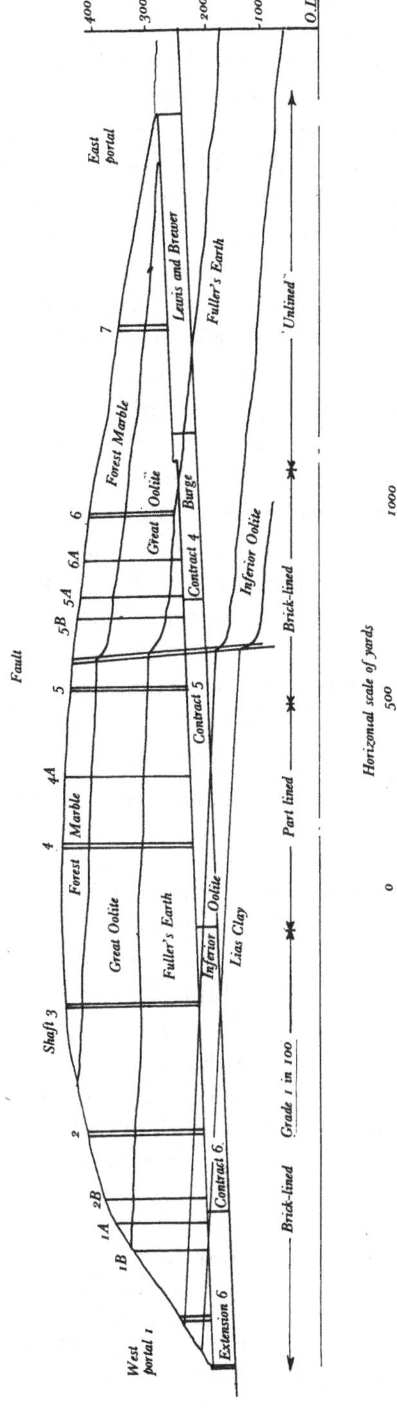

Fig. 6. Box Tunnel: longitudinal section

25 foot shaft into the 30 foot wide arched tunnel.

The downhill grade from east to west of 1 in 100 provoked exaggerated agitation and some wild arithmetic by the critics on the speed of possible runaway trains, so Brunel paid a visit to the Whitstable–Canterbury Railway where a tunnel with a grade of 1 in 100 had been worked by trains since 1826. This tunnel was recently abandoned but has since drawn attention to itself by causing settlement to a building of Kent University.

Although there are several contract numbers on the section drawing, there were only two in the event. During 1837 Brewer of Box and Lewis of Bath were awarded the contract for half a mile of tunnel from the east end and then George Burge obtained a contract for the rest dated and signed by him and Brunel on 30th June, 1838. This contract was engrossed by scribes in all the fashionable whirling penmanship, while the first word on each fresh page is several times the height of the script and illuminated with complex scrolls. The contract is handwritten on thick paper about 28 inches long and 23 inches wide and covers several pages of contract conditions in the best legal tradition, with no paragraphs and little punctuation—so very different from the modern standard conditions. The specification has separate headings and is more modern in its wording, but the bill of quantities would distress a modern quantity surveyor as it consists only of seven unnumbered items.

"Driving the tunnel exclusive of masonry or brickwork	per cubic yard	10/6	Ten shillings and sixpence (signed) George Burge.
"Brickwork set in mortar and including all erection of centres scaffolding jointing excavating etc.	per cubic yard	2/3/6	
"Ditto set in Roman Cement ditto ditto	per cubic yard	2/8/6	
"Coursed rubble set in mortar ditto ditto	per cubic yard	18/6	
"Ditto ditto set in Roman Cement ditto ditto	per cubic yard.	1/1/6	
"Ashlar set in mortar ditto ditto	per cubic foot	1/0½	
"Ditto set in Roman Cement ditto ditto	per cubic foot	1/2½	

"This contract is to extend to the open cutting on the West Mouth of the Tunnel (signed) George Burge."

Brewer and Lewis's contract for half a mile from the east portal was in the Great Oolite and was also serviced by shaft No. 7. Brunel no doubt judged that the Great Oolite would be self-supporting as the area around Corsham was riddled even then with underground quarries which he could inspect. He also made good use of the material for building the portals of his tunnels. The geologist W. J. Arkell (1933) writes:

> "The Bath Freestone is still extensively mined in the district south west of Corsham. The excellent qualities are said to have first been generally realised when it had been pierced for the Box Tunnel. Previously the demand had been only local, but after the coming of the railway, Bath Stone soon began to be sent to all parts of the country . . . ."

There are three coloured cross-sections on the contract drawings but no written descriptions. One shows an unlined tunnel with a Gothic arch 36 feet above rail level and minutely dimensioned with many radii to give a true shape. This is the section which was used in the Great Oolite at the eastern contract but, after some bitter winters, falls of rock occurred and part of it was lined by turning a flattish brick arch and filling the space above with faggots. This method was repeated as late as 1894 as there is a coloured drawing so dated and inserted with old drawings.

Blasting by gunpowder was used in the rock and possibly in the hard clays and, in those days and for many years, the holes for blasting were made by hand-held steel drills. The hammerman would strike while the holder-up would rotate the drill after each stroke. Relays of small boys called tool carriers, usually sons of miners, would run to and fro to take the drills to the blacksmith for constant re-sharpening after about 2 feet of drilling.

Simms (1844) in *Practical Tunnelling* writes:

> "Owing to the constant flow of water from the numerous fissures in the rock extensive pumping machinery was needed. From November 1837 to July 1838 work was suspended as the water gained on the steam pumps employed filling the portion then completed and reaching 56 feet in the shafts . . . A second pump of 50 H.P. was then applied and water pumped out at the rate of 32,000 hogsheads per day."

A hogshead of beer is said to contain 54 gallons so the pump discharge would be about 1,200 gallons per minute, presumably from all the pumps, and would not be the rate of inflow. From the date this must have been from shaft No. 7.

Shaft 3 is the deepest, being 290 feet to rail level, but shafts 2, 3 and 4 are nearly as deep. At the Box Tunnel the winding from the

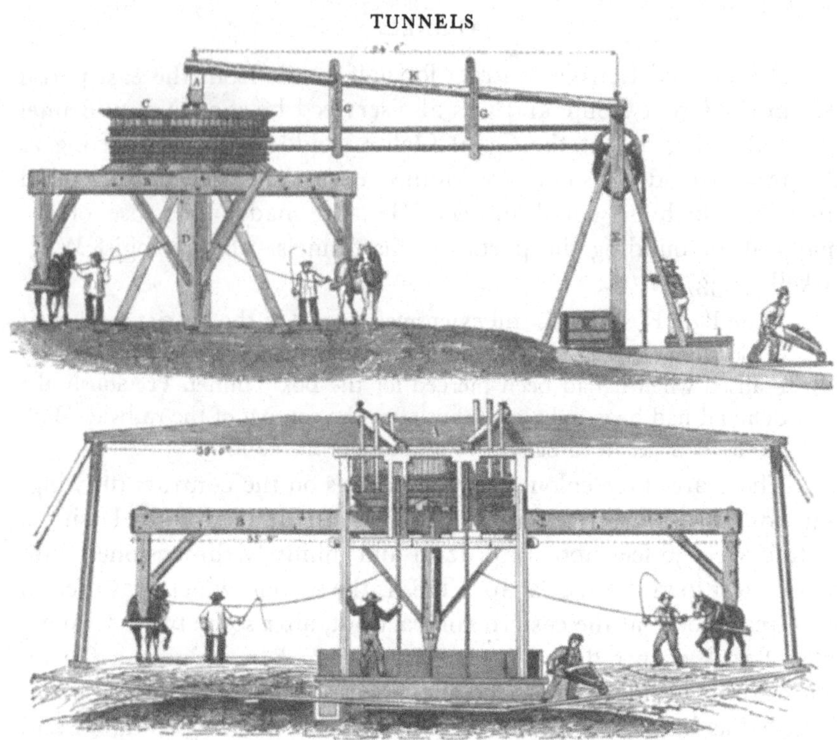

Fig. 7. Box Tunnel: horse gins

shafts was done by horse gins. Fig. 7 shows an example. If two horses were to walk round a circle of 90 feet circumference and if the drum were 30 feet in circumference and if the horses were to walk at 2 miles an hour, a bucket could be wound out of a 260 foot shaft at about 60 feet per minute, which is close to the rate of a 3 ton steam crane. In addition there were two ropes on the double drum so that one skip ascended while another descended. For lack of detail it is possible to assume that the horse gin could wind as quickly as the miners could excavate and timber a length. On the Thames Tunnel the winding was by steam and the mechanisation was in advance of its time, with the Brunels in full charge. The railway tunnels were carried out by contractors, so presumably Brunel would leave the choice of plant to them so as not to relieve them of their responsibilities under the contract. If steam plant had been used on all the shafts it is possible that as many horses would have been needed to haul the coal over long distances and bad roads as were needed to work the horse gins, which were far from inefficient.

The remaining contracts carried out by Burge were in ground

Fig. 8. Box Tunnel: tunnelling method

which would have needed partial or total temporary support. Details of the actual tunnelling methods used in the Box Tunnel are sparse but much can be deduced from Simms' *Practical Tunnelling*, and it can be assumed to have been by the "English method" which had been perfected 50 years before in the canal building period. Fig. 8, taken from Simms' book, shows the method as used at that time. A heading is driven ahead in which the crown bars are placed with one end resting on the brick lining and the other supported on timber props. Sometimes a bottom heading was driven as well to speed the work. The miners would then work on either side to widen out and would support the ground with poling boards which would be supported at their tops by the bars and wedged off the ground at their lower ends. Another bar could then be dragged and rolled into position to rest on the benching, and in due course would be supported at the leading end by props. When the arch had been supported on bars and props the rest of the length could be dug, as if in the open, under this arched support.

The length and size of the bars would be decided by the Engineer's opinion on the ground load, and larch bars were preferred as they had the reputation of giving audible warning by creaking if the load was excessive so that collapse could be forestalled by extra props. In good ground the poling boards would be sparse, but in bad ground or swelling clays the length would be fully timbered. Swelling clays can exert considerable pressures when exposed and in this method the time to complete a 12 foot length could, according to Simms, take twelve shifts to excavate and four days to brick up the lining. He also wrote that at Box 6 inches were allowed for expansion of the clay between the face of the work and the timbers. When a length was completed the miners would then withdraw and start another length in another face while the bricklayers built the lining. The arch would be turned on timber drums with miners in attendance to strike the timbers and remove the bars as each setting of boards was supported by the brickwork. As the tunnel face was 30 feet high, heavy strutting would be needed to support it.

Simms tells us that on the Box Tunnel 5,500 bricks were needed per foot of advance and that they were made locally by 100 men. Rolt (1957) records 1 ton of gunpowder and 1 ton of candles every week while 30,000,000 bricks were carted to the site by 100 horses. In all, 247,000 cubic yards of spoil were wound out of these deep shafts and disposed of. The peak work force was 4,000 men and 300 horses, so the much-used phrase of carrying out work like a military operation seems appropriate. As an analogy to give an idea of the magnitude of the work, five battalions and several squadrons of cavalry would need to be quartered, fed and watered.

The tunnel between shafts 3 and 4 would have been driven with Fuller's Earth above and Inferior Oolite below. A cross-section shows the brick lining as 30 feet in clear width inside and 25 feet from rail to crown. The upper part is 2 feet 7½ inches thick and the lower is 1 foot 6 inches thick. Around shaft 2 the Inferior Oolite was at the top and Lias Clay below. The cross-section for this length is the reverse with the thinner lining above. No doubt the intention was to adjust the thicknesses to the dip of the strata as the ground became exposed in each length.

The tunnel was ready to be opened by June 1841, "after Brunel and his Resident Engineer had forced the pace". Engineers can propel contractors only by threat of dismissal for lack of diligence, by the hope of extra reward or by fear of non-payment, but in this case it is possible that the strong personality of Brunel also had an effect.

During all this time Brunel was actively pushing on with the building of the Cheltenham and Great Western Union Railway, which branched off the main line at Kemble. Here he had to placate Squire Gordon of Kemble by driving a tunnel a quarter of a mile long so that the railway should not be visible from Kemble House. Further along the Sapperton Tunnel had to be driven through the scarp of the Cotswold Hills. In April 1837 the Directors instructed Mr Brunel "to proceed immediately to lay down the line of the railway definitely . . . and to offer contracts . . . for the sinking of the shafts in the Sapperton Tunnel at the earliest period". The document went on to express "some hesitation in making a call on the Proprietors pending severe pressure on the money market but would enable them to effect all which could be prudently accomplished before the winter months". Then follows a page of lamentation on the state of affairs, "but with hopes that the return of a more healthy state of trade and consequently of confidence and credit will restore both the means and desire of embanking capital in works of this description" *(PLB, Apr., 1837)*. This has a familiar ring. Brunel replied in May:

> "The preliminary shafts of the (Sapperton) tunnel have been carried to a considerable depth and are still in progress, hitherto the results of the examination of the strata have proved more satisfactory than there was originally reason to hope." *(PLB, 3 May, 1837)*

Then followed a long statement of the Engineer's endeavour to reduce the cost of the works without impairing their efficiency. This shaft sinking coincided with that at Box and again shows Brunel's admirable sensitivity to ground conditions. These shafts were 18 feet in inside diameter with a lining 2 feet 6 inches thick.

In November he wrote *(PLB, 5 Nov., 1838)* to Wm Crawford:

> "Dear Sir,
>
> "I want immediately a section of the ground over the Sapperton Tunnel and also a continuation of the section of Contract No. 1, Ciren[r] northwards to the P.P. road, the boundary of Kemble parish—this I believe in existence send each as quickly as possible.
>
> "Yrs very truly
> I. K. Brunel."

Then a week later Brunel wrote a long flowing letter to Lawrence which illuminates the squabble with Squire Gordon of Kemble, who had to be bought off to the tune of £7,500. In one of his long sentences Brunel wrote:

> ". . . there must be one or two conditions insisted upon which must not be yielded on any consideration . . . Pray beg the Directors to be positive and unyielding with Gordon . . . I send you the specification of

the Sapperton Tunnel Contracts and one of the drawings which Richardson left behind last night by mistake I perceive Mr. Gordon's object in attending the Board to take advantage of some chance omissions . . . *One* man has always the advantage over a number . . . if he is cunning which Gordon certainly is—he seems to be really savage with you." *(PLB, 12 Nov., 1838)*

In 1845 Brunel wrote a short report to "The Directors of the Great Western & Cheltenham U Railway". In it he stated that "the works upon the Cheltenham Railway are now drawing to a close and the Tunnel completed" *(PLB, 8 Feb., 1845)*. So the Sapperton Tunnel, which started in 1837, seems to have taken eight years to complete its length of 1 mile 100 yards. Possibly a certain amount of stop–go was adopted to fit in with the rest of the work and the financial stringencies mentioned. Also this progress would allow for gangs of miners to become free from the Box Tunnel in 1841.

Another tunnel which obtained notoriety in 1852 was the Mickleton Tunnel through the escarpment of the north Cotswold Hills near Chipping Camden. Rolt (1957) has given a colourful account of how the contractor, Marchant, stopped work and occupied the workings by organising what, today, would be called a "sit-in". Brunel promptly organised his own private army from his other works and, after a rough encounter, drove off the enemy (before the police could arrive) and took complete control. This tells us little about the actual tunnelling, but this example of organising the opposite of a "flying picket" tells us a great deal about Brunel.

Thus Isambard Brunel's major tunnelling works were completed in the same year that his father was able to complete the Thames Tunnel and to pass his grandson through the junction length. The remaining tunnels on extensions of the Great Western Railway do not seem to have caused so much interest. Railway tunnelling had become so widespread that with so much experience crises were prevented.

Brunel's son (I. Brunel, 1870) records that:

> "Mr. Brunel devoted much time to a careful investigation of the Severn to determine the most suitable place for crossing."

This took the form of steam ferries carrying passengers and freight between large timber jetties on either shore. These jetties were connected to pontoons held between piles and the pontoons rose and fell with the tide which has a very high range. The method was one of the inventions of his famous father. Perhaps Brunel doubted whether the financial nerve of his Directors would have been strong enough to contemplate a shield driven tunnel under the River Severn.

III

Clifton Suspension Bridge

SIR ALFRED PUGSLEY

Clifton Suspension Bridge was a long time in the making. From 1830 until his death in 1859 the work was in the hands of Brunel, although actual construction ceased in 1842 and the original chains, made but not erected, were sold in 1849. From 1860 to 1864 the bridge was completed in its present form by John Hawkshaw and William Henry Barlow. In the 100 years that have followed the bridge has been maintained by a private company and by Trustees, advised since 1910 by Howard Humphreys and Sons as consulting engineers. It is convenient to discuss the bridge structure in terms of these three periods—1830–60, 1860–64 and 1864–1974—although of course the work of the periods is interconnected.

Brunel came to the problem of bridging the Avon Gorge at the age of 24, and with no specialist experience of suspension bridges beyond, when in his teens, that of working with his father when the latter was designing two small suspension bridges (each of 132 feet span) on the Ile de Bourbon in the Indian Ocean. But he had a knowledge of French theory in this field and a willingness to learn from earlier examples; moreover, he had the confidence and energy of youth and was stirred by the magnificence of the site.

From the first, Brunel dismissed any thought of spanning the gorge by other than a suspension bridge, and that one of a single span. He realised that this would mean a span greater than ever before achieved—Telford's bridge over the Menai Straits, the longest then extant, had a central span of 580 feet—but he was aware that his father, at Saltash, and Telford, at Runcorn, had contemplated much greater spans. In November 1829 Brunel submitted four designs, with spans ranging from 760 feet to 1,160 feet, for four possible adjacent sites. Of these Brunel himself favoured, largely on aesthetic grounds, that with a span of 980 feet which, emerging from a tunnel on the Clifton side, would "burst upon the splendid scene" and "would altogether have formed a work perfectly unique and the grandeur of which would have been consistent with the situation" (Porter Goff, 1974). In this situation the chains could enter the rock direct on both sides, saving towers and land spans alike.

In the event, due mainly to criticisms by Telford, who by then was 73 and rendered cautious by oscillation troubles on his Menai Bridge, a second set of designs was called for in December 1830, when Brunel meekly submitted a scheme involving a span of only 600 feet. This, after some adjustments to satisfy the assessors Davies Gilbert and John Seaward,[24] won the competition against all twelve other competitors. But within a few months, when he came to detailed design work, he had persuaded the Bridge Company to agree to a span of 702 feet, as exists at present.[25]

Paramount in the minds of suspension bridge designers at this time were problems concerning the assessment of loading on such bridges, the strength of chains and the danger of damage due to wind-excited oscillations; and all were thought likely to become more troublesome with increasing span. Although opinions as to the loading for which long span bridges should be designed differed widely, all were agreed that the greatest loading would arise from a crowd of people. When the Runcorn project was assessed in 1818 estimates of loading varied from 7½ pounds per square foot to 24½ pounds per square foot, but by 1830 J. M. Rendel was using 45 pounds per square foot and Brunel 100 pounds per square foot. There were similar differences about the strength of wrought iron chains. Although it was common knowledge that a small bar of wrought iron, when tested in tension, yielded at about 13 tons per square inch and broke at about 23 tons per square inch, such was the concern about the properties of large welded or forged links that, among the potential designers of the Clifton bridge, "safe" values of design stress of 4–10 tons per square inch were proposed. Brunel himself became more conservative as time went on: in 1829 he proposed 8 tons per square inch, in 1830 6½ tons per square inch and by 1838 he adopted 5 tons per square inch. Davies Gilbert, faced as an assessor with this situation, compared the various designs on the basis of a loading of 84 pounds per square foot and a safe tensile stress of 5½ tons per square inch; these figures, so far as the chains were concerned, favoured Brunel's design (Porter Goff, 1974).

Great concern was felt about the danger of damage due to wind-excited oscillations. It is intriguing to find that as early as 13th February, 1830, some lay directors of the Bridge Company waited upon Brunel's father to seek assurances on this matter. Sir Marc says:

> "I explained to them how the lateral agitation may be prevented, and how the effects of the wind might be counteracted." (Beamish, 1862)

He had in mind the system of staying catenary cables below the bridge deck that he had successfully used at the Ile de Bourbon. Brunel himself had from the first thought to use some such cables, and assumed to the end that these would be necessary at Clifton. But he realised that for the long span at Clifton any "inverted" catenary cables for staying the bridge deck from below would, with temperature changes, oppose the main chains and give rise to severe internal stresses. To obviate this and produce a definite initial tautness, he planned to fix each of his two inverted catenaries at one end only, the other end being attached to a lever system and loaded by a dead weight there.[26]

Brunel was formally appointed to design and construct the bridge on 26th March, 1831, but his father, anticipating this in December 1830, and:

> ". . . bringing the experience of his Bourbon bridges to bear upon this new project, at once devoted his energies to render the mechanical arrangements for the Clifton bridge complete. On 12th January, 1831, he records in his journal: 'Devised this day, for Isambard's bridge, a new mode of carrying the heads of the chains, and sent him a drawing of it to Manchester' . . . Through the months of February, March and April, there were few days that he was not engaged upon that interesting work." (Beamish, 1862).

It is very evident that the father who trained his son at the Thames Tunnel was determined to give him a flying start on his first independent project.

Brunel now set to work on the precise siting and design of the bridge foundations and towers. He was committed to the most southerly site of his various proposals: that joining two almost level rock platforms, to the west near the Iron Age camp at Burwalls and to the east on the limestone promontory just to the south of the Observatory Hill. He was also restricted to a clear span of only 600 feet or thereabouts. In the event, he placed the Clifton Tower boldly and strikingly close to the precipitous edge of the Clifton platform, and brought the Leigh Woods Tower off the Burwalls platform on to a vast masonry structure rising from the lower limestone slopes there and backing on to the natural platform above. By so doing his final span of 702 feet was achieved at the cost, as he pointed out at the time, of much extra expense on the Leigh Woods foundations. Unfortunately, no detailed drawings of these foundations exist, but there is a passing reference in the Brunel papers of the Clifton Suspension Bridge Trust *(CSBT, 1838)* to the incorporation of iron land ties, presumably to bind the masonry abutment to the rock slope behind it.

Special investigations were made in 1969 to check the safety of this great abutment. Deep drillings around it showed that the structure rests partly on the carboniferous limestone and partly on Triassic conglomerate and breccia subsequently formed in gullies in the limestone slope. Electrical resistance measurements between the drilled holes led to the conclusion that the rock beneath the abutment was sound and free of any large cavities such as are known to occur in the gorge cliffs. These investigations were made partly because of known "slips" elsewhere in the gorge and partly because of an increase in water seepage through the rocks in the vicinity. The reassuring results are a tribute, after 140 years, to Brunel's judgement in his siting and construction of the Leigh Woods abutment.[27]

The towers themselves, apart from their capping and the absence of their planned Egyptian ornamentation, are much as Brunel left them. The loading on them, however, is a little more severe than he intended, partly due to the greater weight of the present bridge and partly because of a change in the inclination of the land chains.

We come now to the suspension chains themselves. Brunel was keen from the start that there should be only two chains, one each side of the roadway, and each made of wrought iron links as long as practicable. The object of using long links was to minimise the weight and cost of the joints, and the desire for a single chain on each side was to avoid the uncertainty in the sharing of the load between the chains should more than one be adopted. As regards the links and their joints, Brunel was able to overcome the initial objections of Gilbert and Seaward, but as regards the number of chains Gilbert was adamant and Brunel was forced to plan for four chains in all, two each side.

Here Gilbert was aiming for a degree of redundancy in the structure to ensure that the structure should be, in modern parlance, a "fail-safe" one. He raised this idea in his first report: "all the parts" should be "sufficiently divided so as to afford a power more than equal to supporting the bridge notwithstanding a failure in any one part, and also afford ample means for replacing any part that has failed" *(CSBT, 17 Mar., 1831)*. The young Brunel evidently took this lesson to heart—as many succeeding generations of structural engineers have had to do—and applied it to many parts of his bridges at Clifton and elsewhere.

The overall efficiency of a suspension bridge is considerably affected by the amount of the dip given to its main catenaries. Telford at Menai chose a ratio of dip to span of 1 to 13.5; this is shallow in modern terms as cost investigations now give optimum values of 1 to 10 or 1 to 11. Brunel adopted 1 to 10 and so not only achieved some economy but also reduced the pulls on the anchorages. The land chains or backstays he planned to reflect in slope the chains in the main span and to be "tied down" to the ground by suspension rods matching those in the span. By this means, and by bringing the main chains down to the level of the bridge deck at its centre, Brunel hoped to add to the resistance of the structure to oscillations due to wind or traffic. This last feature was objected to by Gilbert, who asked him to raise the chains so that they were visibly clear of the deck structure. Gilbert seems to have insisted on this mainly for aesthetic reasons, but the change had the technical advantage of leaving the chains free to swing in their own planes without pulling on the bridge deck and so causing (as on some later suspension bridges) attachment failures there.

The chain links as first designed by Brunel were at the then practicable limit of 12 feet long, but by 1832 he was able to plan for 18 feet and by 1838 for 20 foot links; and it was these that were actually made and later sold for use at Saltash. In his first designs Brunel intended to construct each of his two chains of 16 links made of plates $7\frac{3}{4}$ inches by 1 inch in cross-section, with special lugs welded to each end. The links were to be joined directly by large pins passing through their overlapping end lugs. By this means he planned to save the short interconnecting links adopted by his predecessors. By 1840, with the adoption of four chains instead of two, his designs show links 7 inches by 1 inch in cross-section and varying in number from eighteen at the centre of the span to twenty at the towers.[28] In all his designs the "shoulders" provided at the back of his lugs were to permit of the removal of a link by cramping adjoining links together by grips over the shoulders—an instance of Brunel's response to Gilbert's replacement principle.

The proportions of the lugs adopted by Brunel (Fig. 9) were a matter of interest to many engineers of his time, as they are indeed today. His work in this field was discussed at length soon after his death by his assistant Brereton at a civil engineering meeting (Brereton, 1870) and met general approval, which was later substantiated by further work on the problem by Hawkshaw for his Charing Cross railway bridge. An examination of Brunel's lugs in terms of the results of extensive aeronautical work on the steel lugs for the bracing members of biplanes during the 1930s again pointed to the merits of Brunel's designs for the Clifton bridge.

The manufacture of these lug ends to the links was a matter of some difficulty at the time. Telford's smaller links at Menai had lug ends welded to the link bars, but Gilbert and Seaward, evidently anxious about the reliability of such welded ends, said in their 1831 report "the greatest attention" should be "paid to the preparing of the links of the great chains, the iron should be entirely made under the great hammer and when carefully forged to the shape should be proved to 10 to 12 tons per square inch before the holes are bored"[29] *(CSBT, 17 Mar., 1831)*. But by 1840 Brunel had decided on welded ends—the successful forging of large integral lug ends awaited Howard's patent process (1845) using two sets of rollers at right angles —and to check their reliability by proof testing every link in its finished state, loading through pins to a stress of 9 tons per square inch in the parallel portion of the link (Brunel's specification, *BS, 1840*).

Each chain in Brunel's 1830 design passed over rounded saddles at the tops of the towers and thence down to fixed saddles at ground level, and from these to the anchorages. In these matters Brunel seems to have been following the Menai design, but by 1840 he had adopted his father's design of roller saddles at the tower tops, arranged to carry two chains, one above the other on its own roller system, at each side of the bridge.[30] Saddles on rollers were now planned at ground level and the chain anchorages were to be an improved version of those at Menai. The eighteen links of each chain at the land saddles were to pass down into a narrow inclined tunnel to two cast iron anchorage frames, one above the other, some 20 feet apart. Each frame was of wedge shape designed to pull against sloping shoulders of rock formed by a local enlargement of the tunnel. At each frame the links passed through a "comb" of plates integral with the casting and were secured against the comb by a heavy cotter through all the links. Between the upper and lower anchorages, the number of links in the chain was reduced from eighteen to sixteen (Brunel's drawings, *BD, 1840*).

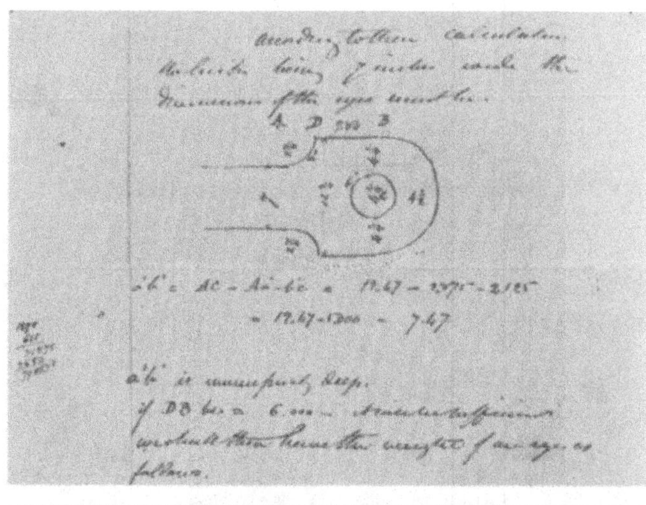

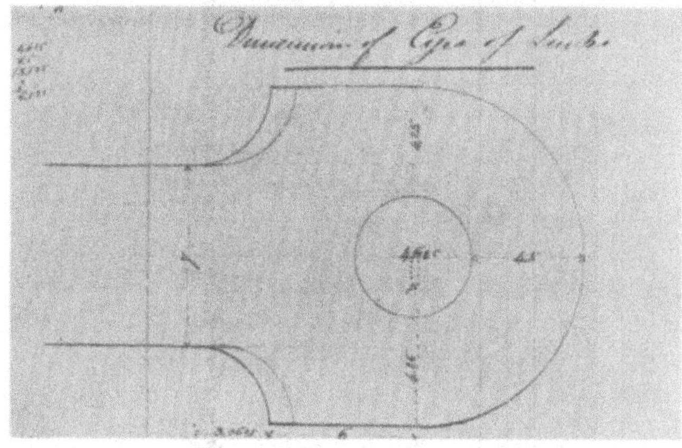

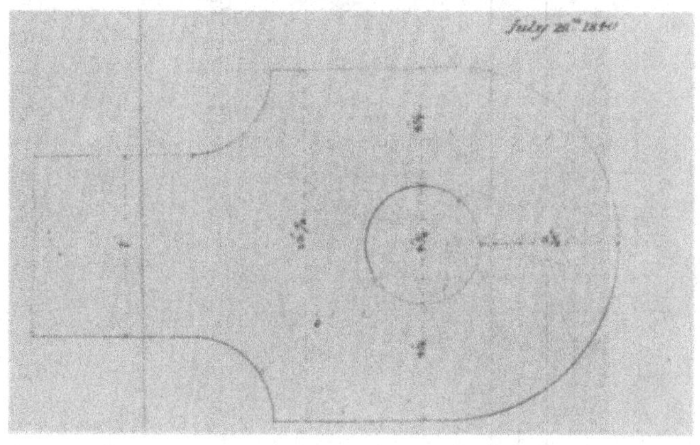

Fig. 9. Clifton bridge: development of link lugs

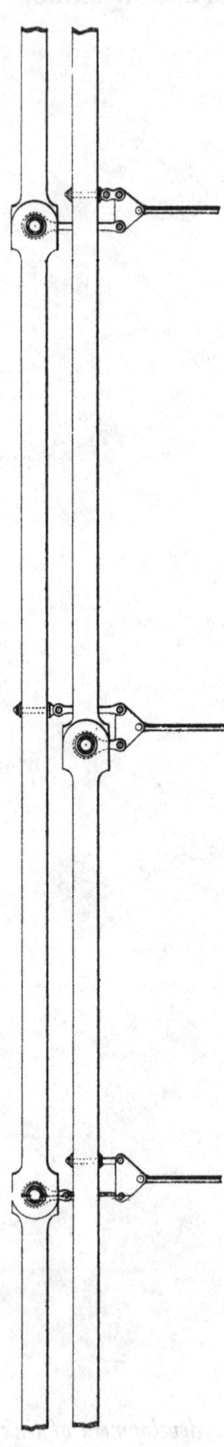

Fig. 10. Hungerford Bridge: link and suspension rod arrangement

In August 1840 Brunel reported, "with regard to the construction of the chains, a satisfactory contract . . . has been entered into with Messrs. Carne & Vivian of Hayle, Cornwall, who are now engaged in their formation" and "the tunnels or shafts for fastening the main chains . . . are now under execution" *(CSBT, 6 Aug., 1840)*. But in December 1842 Brunel was asked to stop work on the chain iron, and in July 1849 he was asked to sell it. The inclined anchorage tunnels he cut were never used and no sign of them now remains.

Coming now to the suspension rods or hangers, Brunel at first intended to space these at 12 foot intervals throughout the length of the chains, backstays included. Each rod was to hang direct from the pin through the lug ends of adjoining 12 foot long links, and was to have a strength capable, in the suspended span, of carrying the dead weight of its share of the deck plus a superimposed load of 100 pounds per square foot thereon, at a tensile stress of only one half of that allowed elsewhere. This special margin was provided, in agreement with Gilbert, to allow for the possibility of an adjoining hanger breaking.

When, however, Brunel found he was able to obtain links up to 20 feet in length, and at the same time was pressed into using two chains at each side of the bridge, he kept the hanger spacing at 10 feet, but to do so proposed an arrangement for attaching each hanger to both chains. This was to be done by fastening each hanger at its top to a small link plate, which was itself suspended by two short rods equally spaced about the main rod, hanging one from the main pin in one chain and the other from a strap fastened over the links of the other chain at their mid-point. Brunel later used this arrangement for sharing the load between the chains at Hungerford (Fig. 10). As noted by Porter Goff (1974), it was a better one than that used by Telford at Menai, but introduced substantial bending stresses in the links themselves, an unexpected scheme for so sensitive a designer as Brunel. Yet neither he nor Gilbert seem at any time to have mentioned or assessed these bending stresses, which would have added some 40 per cent to the direct tensile stresses in the links.

One of the problems involved in the design and erection of the hangers of a suspension bridge is that of ensuring that, in so highly redundant a structure, each hanger carries its proper share of the deck load. There is here a problem seemingly worse than that of tuning the wires in a harp, for part of the "frame" is a flexible chain. Two possibilities present themselves: to calculate the precise length of each hanger to satisfy, when loaded, the intended geometry of the chains and deck, and to trust that after all is assembled the partition of the dead loading will be even among the hangers; or to fit the suspension rods to approximate lengths only and in the final assembly to adjust these lengths (by turnbuckles) until each rod carries the same load—a tuning process. Brunel, from his calculations, decided to adopt the first procedure, and most suspension bridge designers since have followed him. All are aware that a proper distribution of the loads among the hangers cannot be accurately achieved. Airship and biplane designers, using "open panels" in their framed structures cross-braced by cables or thin rods, were forced to adopt the second procedure, largely because of the special redundancies introduced by the cross-bracing. But perfect "tuning" of the bracing cables was never possible and there were almost always some slack wires, and frequent "re-tuning" was necessary.

An unexpected feature of the suspension rods appears in the 1840 drawings. The $2\frac{1}{4}$ inch diameter rods had forged fork ends welded to their lower ends, and these were to be pinned to lugs on the longitudinal girders of the deck with $2\frac{3}{4}$ inch diameter pins inserted parallel to the girders. By so doing, Brunel had apparently failed to appreciate fully Telford's experience at Menai, where suspension rods pinned in this way failed, partly because of the local bending thus induced in the rods when the main chains, excited by traffic or winds, swung like pendulums in their own planes.

The suspension rods, at their lower ends, were to be attached to timber cross-beams on which the deck was to be constructed. The arrangement Brunel seems to have had in mind in his first design is illustrated in Fig. 11(a), and is like that he used later at Hungerford. The cross-beams, about 12 inches deep and 6 inches wide, extend beyond the chains to carry footpaths, and were to be placed at 6 foot intervals so that every other one was carried by the suspension rods. Longitudinal "whole timbers", about 12 inches by 12 inches in section, rested on the cross-beams and all were secured together by extra longitudinal timbers, about 12 inches by 6 inches in section, running beneath the cross-beams. Long vertical bolts, passing through all three timbers—upper and lower members and the cross-beams—held all together and carried lugs to which the hangers were attached. Timber planking, placed longitudinally on the cross-beams, covered by about 2 inches of gravel, provided the surface for the road and footpath.

The evidence for this reconstruction of Brunel's first scheme is not good. It is based mainly on sketches—not always identifiable—in the

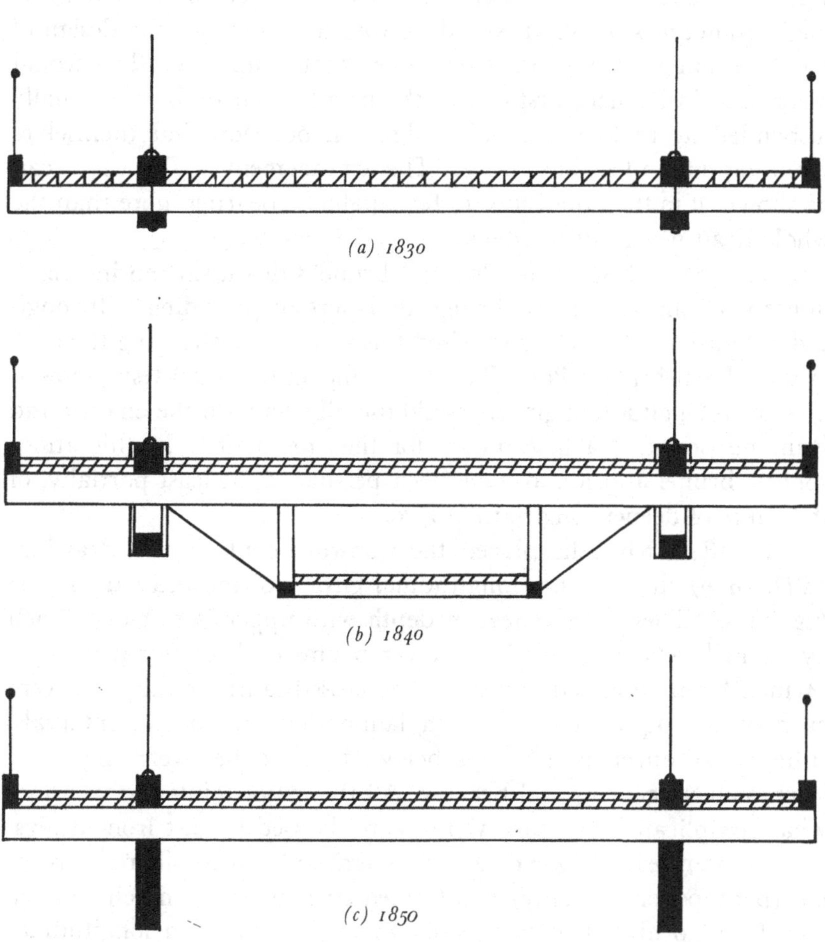

Fig. 11. Clifton bridge: development of deck design

Bristol collection and on the deck structure he used at Hungerford. But it is clear that at this time (1830) he had no intention of providing a deep longitudinal girder system and was not concerned with the stiffness of his deck structure.

He was, however, concerned about its strength, which he based on a principle enunciated, with Gilbert, as follows: that since the hangers might sometimes break, it would be wise to assume for the design of the deck that any three might be broken at the same time. This would mean that in Brunel's first design the longitudinal timbers, normally suspended at 12 foot intervals, might on occasion find themselves spanning not 12 feet but 48 feet. The arrangement in Fig. 11(a) can be shown, if in this condition, to be capable of bearing more than the whole dead weight of the deck.

The interval between 1830 and Brunel's death saw an increased interest among suspension bridge designers in providing stiff longitudinal girders. Rendel emphasised this when reconstructing the deck of the Menai bridge; Peter Barlow, by means of model tests, showed how deep longitudinal girders could usefully act with the chains; and Rankine provided a first theory for this interaction. In this atmosphere, Brunel appears to have been persuaded, at least partially, of the value of deeper longitudinal girders.

By 1840, when he placed the ironwork contract, his drawings *(BD, 1840)* show timber longitudinal girders of the Pratt truss type Fig. 11(b). These were 4 feet in depth with upper booms of 12 inch by 12 inch whole timbers and lower booms each of two 9 inch by 4½ inch laminations side by side. The cross-beams for the deck were each of two 14 inch by 4½ inch laminations at 10 foot intervals, stiffened by queen post trusses below; these trusses were stayed by interconnecting struts and bracing. All the major joints in this somewhat complicated structure were neatly housed in cast iron fittings.

In later years, however, Brunel seems to have considered replacing (perhaps for economy) this framed structure by a much simpler one. In 1850 his calculation books show (Fig. 11(c)) a longitudinal system consisting of an upper timber 12 inches by 5 inches above the cross-beams with a solid timber beam, built up in laminations, 36 inches deep by 5 inches wide below *(CB, 1850)*.

There is some evidence that at a still later stage Brunel reverted to his trussed type of longitudinal girder, but with the depth increased to 6 feet (W. H. Barlow, 1867). As compared with his first design, these later schemes would have increased the flexural stiffness of the deck by up to 100 times, but such a stiffness would still have been much lower than would have arisen from a direct application of Rankine's theory. No thought was given by Brunel or his contemporaries to the torsional stiffness of the deck, although all his schemes provided horizontal cross-bracing by wrought iron bars placed diagonally between the cross-beams.

Brunel's schemes for the Clifton bridge, spread as they were over nearly 30 years and always bedevilled by lack of funds, naturally did not go unchallenged. Early in 1831 the Trustees, sensitive as to the appearance and cost of the "Gothic" towers, asked him to consult a Mr Rickman on the matter. Brunel readily agreed to do so and as a result reported that he and Mr Rickman agreed that the Gothic design "would incur an expense far beyond the intent or means of the Trustees", and instead put forward his "Egyptian" scheme, which at once became popular. In 1835 a Mr West, as a result of his travels in Europe, where he saw the just-completed Freiburg bridge, wrote asking the Trustees to consider using wire cables instead of chains. Again Brunel—though not so readily—had to see Mr West, but reported firmly against the use of wire cables (not a bad decision at a time when one recalls the corrosion troubles that beset many early cable bridges). The Trustees, who evidently had some sympathy with Mr West, formally thanked him and awarded him 30 guineas for his trouble. In 1837 the Trustees received a letter from Thomas Motley proposing a new suspension system (called by him an inverted bracket system, the forerunner of the modern cable-stayed girder bridge) and offering to show them a large model of the scheme. This time five of the Trustees themselves went with Brunel, and "some explanation on the properties and merits of the model was given by Mr. Motley and which occasioned many questions to be put to Mr. Brunel; the result of which was that Mr. Brunel was convinced that the principle of Mr. Motley's design was objectionable" *(CSBT, 13 Sept., 1837)*. The report, perhaps significantly, is by Mr Jones, a Trustee, but nothing more was done about the Motley scheme, although Motley himself later built an excellent small bridge on his plan over the Avon near Bath. Brunel was still, however, to be badgered by outside would-be designers, for only the next year a Mr Protheroe "submitted for the consideration of the committee whether it would not be desirable to obtain Mr Brunel's opinion on Mr Dredge's patent for tapering chains" *(CSBT, 31 Mar., 1838)*. Mr Dredge, who later built a small bridge—still extant—over the Avon in Bath on his principle, had powerful supporters and even sought the influence of Lord Melbourne, then Prime Minister, so it was not surprising that the committee "resolved that Mr. Brunel was requested to meet this committee on the subject at a time most convenient to himself". But by now Brunel, 32 years old, was heavily engaged with the Great Western Railway, and never found a convenient time! Designers of great engineering works involving years in construction have since had similar experiences and have had to brush aside, in one way or another, similar onslaughts.

So much for the first period, 1830–60, of the life of the bridge, the period during which Brunel was in direct control until his death in 1859. Following his death, and that of his friend and rival Robert Stephenson, the third of this "great triumvirate" of railway engineers, J. Locke, became President of the Institution of Civil Engineers and sought with his senior colleagues for ways of completing the Clifton bridge as a memorial to Brunel. It happened that, by an Act of 1859, the South Eastern Railway was authorised to be extended from London Bridge across the Thames to a new station—Charing Cross—and, to make way for this, Brunel's Hungerford Suspension Bridge was to be dismantled and replaced by the present great railway bridge, to be designed by John Hawkshaw. It thus became possible to use the Hungerford chains at Clifton, and with this thought in mind the Institution called into being a new bridge company with W. H. Barlow—later to become famous for the great arched roof of St Pancras station—and Hawkshaw as engineers. It is often suggested that the bridge thus brought to life in the period 1860–64 was in some ways inferior to that finally intended by Brunel. But let us look at the changes that Barlow and his colleagues actually introduced.

They came to the work with the towers largely completed (but minus, as now, their Egyptian finishes), some anchorage tunnels cut and the chains from Hungerford available. The latter were sufficient for two chains, one above the other, on each side of the roadway, but this arrangement involved hangers at 12 foot intervals, each hanging in part from straps over the mid-point of the 24 foot long links, and so introducing severe bending actions thereon. Barlow and Hawkshaw decided at the outset that such bending actions, acceptable though they were to Brunel, must be avoided, and to achieve this planned to use three chains each side with hangers at 8 foot intervals, each hung direct to a main pin in one of the chains. This necessitated making new links for the uppermost chain, and to reduce the number required they steepened the land chains at each end, bringing the points where they entered the ground closer to the towers and removing any necessity for hangers to the land chains. The use of three instead of two chains each side did not involve any appreciable increase in the total weight of the chains: the number of links per chain was simply reduced to alternately ten and eleven.

The dip of the chains remained at Brunel's 70 feet, but these changes necessitated two other alterations. First, new anchorage tunnels were required and these gave Barlow and Hawkshaw an opportunity for a new design of anchorage. The chains, after passing the land saddles just below ground level, now dip down at about 45 degrees to the horizontal for 60 feet, diverging a little as they do so to pass through separate anchor plates backed by brick arches abutting on solid rock. Secondly, Brunel's tower saddles—also from Hungerford and simpler than his 1840 design for Clifton—had to be increased in height to accommodate the third chain, and the roller base tilted slightly (2 degrees, with the upper edge towards the river) to ensure a balance between the chain forces on either side of the saddle.

All the links in the Hungerford bridge had been tested in tension to an average stress of 10 tons per square inch, and the new links, made by Messrs Cochrane, the contractors under Hawkshaw for the Charing Cross bridge, were similarly tested.

We come now to the other main design change made by Barlow and Hawkshaw: the replacement of a deck structure mainly of timber by one mainly of wrought iron. Brunel's final timber truss longitudinal girders were, in the usage of the new times, replaced by wrought iron plate girders of about the same effective stiffness. There is an unresolved problem about these new girders: Barlow, in his 1867 paper, describes these girders as they were actually made and exist, yet in the drawings appended to the paper are shown not 3 foot deep plate girders but 5 foot deep lattice girders! And these drawings are repeated without comment in Humber's book of 1870. So far as the author can trace, this error—or change of design—has never been commented on, either at the discussion of Barlow's paper or since. It seems possible that the lattice design was prepared to meet the growing inclination to follow Rankine and aim for much stiffer girders, and then at the last moment Barlow reverted to the shallower girders. At the time, 3 foot deep rolled I beams (Matheson, 1873) were just becoming available (an experimental one appeared at the 1851 Exhibition) and Barlow may have thought of using them but turned to the plate girder because of its more reliable strength. However, nothing of these matters has been traced in the literature.

The discussion on Barlow's 1867 paper roamed over a wide field but, after some strange misunderstandings about the steep angle of the backstays (now almost standard practice for bridges of this size), concentrated on the longitudinal girders. In the main, most, led by Vignoles, thought these should have been deeper—at least 5 feet, in fact. But many would now agree that the shallower girders make at least for a better appearance.

We come now to the last of our three periods—that from 1864 until today—to comment on the maintenance of the completed bridge (Fig. 12), now over 100 years old. We have already referred to recent geological investigations into the security of the tower foundations. Few corrosion problems have arisen in the wrought iron structure; the visible structure has suffered little save for the end cross-girders under the deck. These were placed too close to the masonry to be painted or inspected properly, and by 1960 had corroded so badly that they had to be replaced—in the absence of wrought iron—by zinc-coated steel I beams. Small cross-tunnels under the end flaps of the deck have since been cut in the masonry to allow ventilation and inspection at these points.

More serious corrosion occurred in the chains near the bottom of the anchorage tunnels, particularly on the Leigh side, where inadequate ventilation combined with ingress of water due to faulty road drains. Back in 1924 the Trustees resolved to strengthen the chains in these regions and asked Professor Ferrier of Bristol University to consult with Messrs Howard Humphreys and Sons on how best to do this. It was decided to add a new link to each chain as it entered the anchorage arch, to supplement the chains in the anchorage tunnels by open-link chains from below the land saddles outwards into small anchorages cut in the tunnel sides, and then to embed the whole in concrete.[31] This was completed on the Leigh side in 1926 and on the Clifton side in 1939. Professor Andrew Robertson and Colonel E. N. Elgood, also of the University, took part in the Clifton operation.

The timber decking of the bridge consists of 5 inch thick longitudinal baulks side by side, covered by 2½ inch transverse planking. This has of course needed replacement during its long life—only partial up to the Second World War, but complete once since then.

A more insidious deterioration can arise from metal fatigue, and during the past decade, because of the greatly increased traffic over the bridge (amounting at peak commuter times to over 1,000 vehicles per hour in one lane), investigations have been made into regions of stress concentration where fatigue cracks might arise. The joints between the cross-girders and longitudinal girders have received

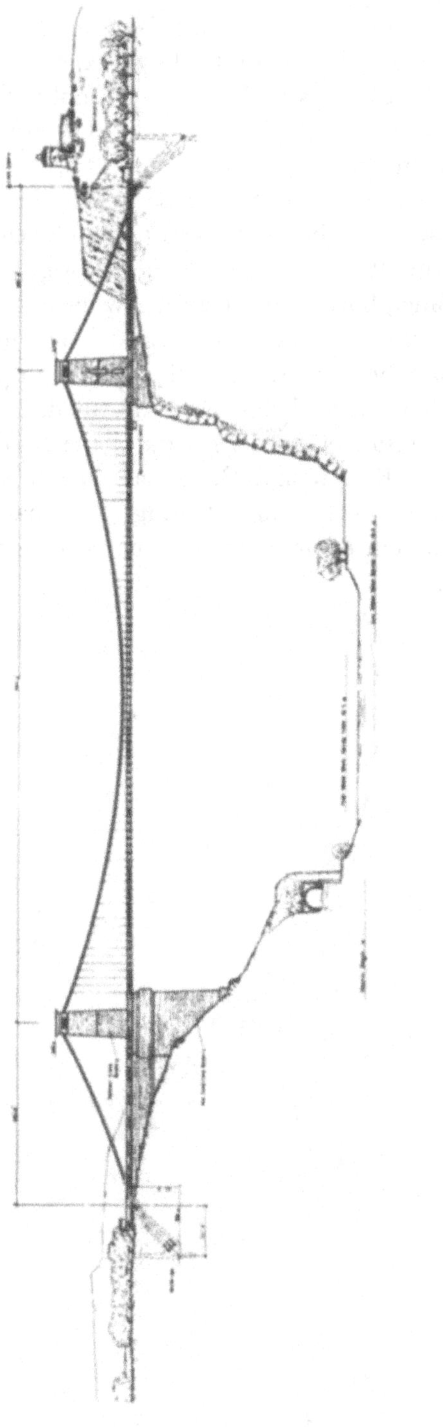

Howard Humphreys & Sons

Fig. 12. Clifton bridge as completed

careful attention and have been modified in detail to reduce any chance of fatigue trouble. Theoretical and experimental studies have since been made by Messrs Howard Humphreys and Sons and Dr Cullimore of Bristol University into the stress concentrations around the pins in the chains, and substantial fluctuations of these tensile stresses have been found to arise at the pins near the towers, where small rotations of the links about the pins occur, particularly in high winds. Fortunately, at these points the original proof testing of each link by Brunel, Barlow and Hawkshaw introduced residual compression stresses that are probably large enough to inhibit the development of fatigue cracks. Special periodic inspections of these critical regions have nevertheless been instituted.

The overall result of Brunel's work at the Avon Gorge, supplemented by that of Barlow and Hawkshaw, has been to bequeath to succeeding generations a beautiful bridge at a magnificent site that has withstood the test of time for over 100 years and may well do so for many more.

IV

Railways

O. S. NOCK

The versatility and volatility of Brunel as a civil engineer was perhaps never more vividly displayed than over his association with railways. Unlike his great contemporaries in the North Country he was not born into the emergent railway age. Whereas the Stephensons, father and son, and Joseph Locke began their lives' work in the hard, primitive world of colliery tram tracks, Brunel was a professionally trained engineer, and showed it when the call came to him to build railways (Nock, 1955). While the men of the North advanced cautiously, step by step, from the practice they had seen from childhood, Brunel submitted the whole prospect to vigorous re-thinking, and laid before the directors of the Great Western Railway what was virtually a new conception of railway construction and travel potentialities. He was certainly entering into this new field of engineering activity on the grand scale, and when appointed on 7th March, 1833, to make the first surveys of the proposed line from Bristol to London, it was the longest railway that had then been contemplated, exceeding by about six miles the London and Birmingham, of which Robert Stephenson was shortly afterwards appointed Engineer in Chief.

Except in its great cuttings and long tunnels, the London and Birmingham was in the continuous line of development from the old colliery railways, but Brunel's new approach led to an altogether grander conception for the Great Western (Nock, 1962). While the construction of the road bed, most of the bridges and the architecture were based on established fundamentals, all provision for the actual running of the trains was looked at afresh, and from a theoretical viewpoint. It was this that led Brunel to features that were to distinguish the Great Western from all other railways: the gauge of the rails and the design of the permanent way. These arose from his conviction that the potentialities of railways for speed should be fully exploited. This was never shown more clearly than in a report to the Directors several years later when the ordering of locomotives was in question:

> "I shall not attempt to argue with those who consider any increase in speed unnecessary. The public will always prefer that conveyance which is the most perfect, and speed within reasonable limits is a material ingredient in perfection in travelling." (I. Brunel, 1870)

In 1835 he wrote a lettei (private letter book, *PLB, 15 Sept., 1835*) to the Directors of the GWR on the subject of the rail gauge, from which the following significant extracts are taken:

> "The resistance from friction is diminished as the proportion of the diameter of the wheel to that of the axle-tree is increased; there are some causes which in practice slightly influence this result, but within the limits of increase which could be required we may consider that practically the resistance from friction will be diminished exactly in the same ratio that the diameter of the wheel is increased; we have therefore the means of materially diminishing this resistance."

This, of course, neglects other forms of resistance, notably that of movement through the atmosphere.

Referring then to the Liverpool and Manchester Railway he continued:

> "The width of the railway being only 4 feet 8 inches between the rails, or about 4 feet 6 inches between the wheels, the body of the carriage, or the platform upon which the luggage is placed, is, of necessity, extended over the tops of the wheels, and a space must also be left for the action of the springs; the carriage and load is raised unnecessarily high, while at the same time the size of the wheel is inconveniently limited.
>
> "If the centre of the gravity of the load could be lowered the motion would be more steady, and one of the causes of wear and tear both in rails and carriages would be diminished.
>
> "By simply widening the rails so that the body of the carriage might be kept entirely within the wheels the centre of gravity might be considerably lowered and at the same time the diameter of the wheels be unlimited."

This principle was in fact used on no more than a few Great Western coaches. Brunel continued:

> "I should propose 6 feet 10 inches to 7 feet as the width of the rails which would, I think, admit of sufficient width of carriages for all purposes. I am not by any means prepared at present to recommend any particular size of wheel or even any great increase of the present dimensions. I believe they will be materially increased, but my great

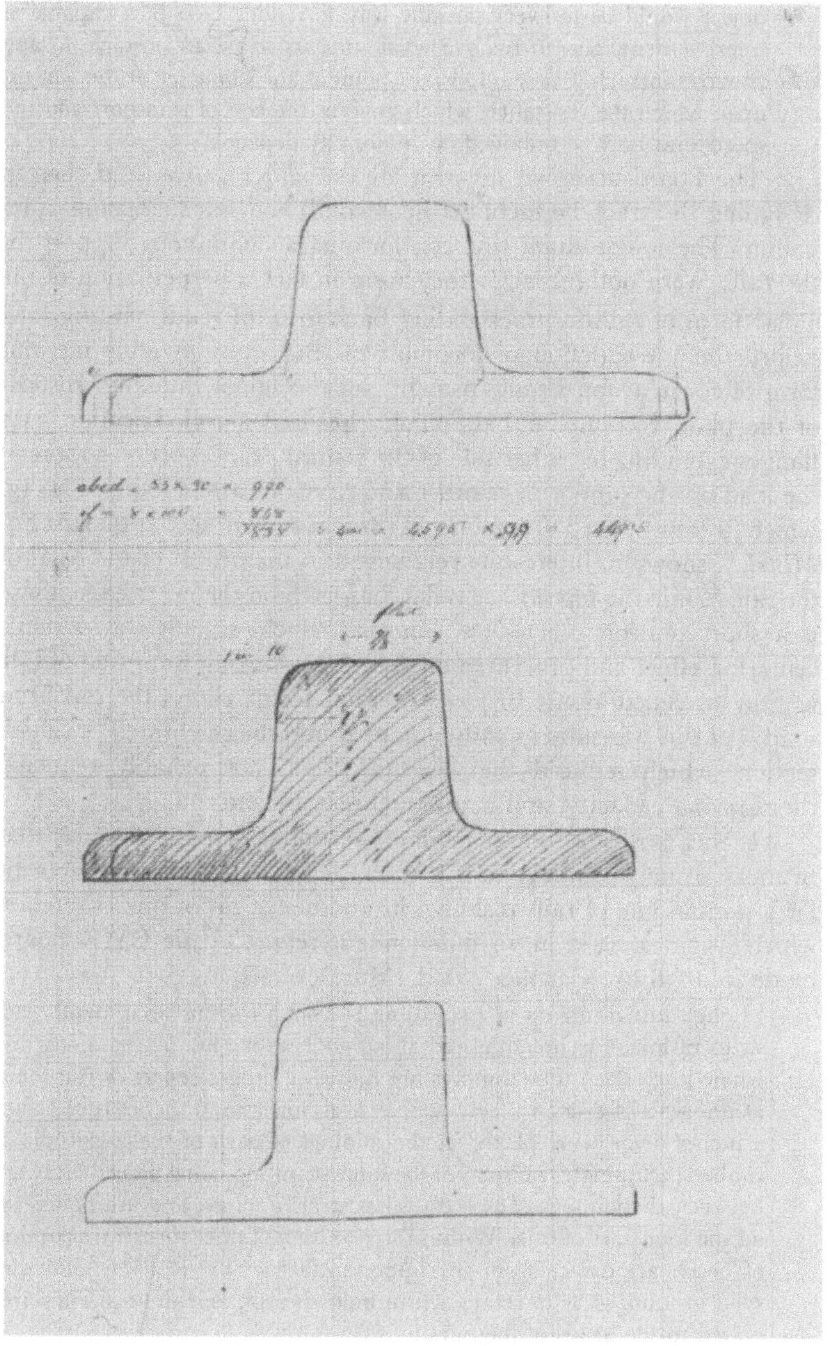

Fig. 13. GWR: solid rail section

object would be in every possible way to render each part capable of improvement, and to remove what appears to be an obstacle to any progress in such a very important point as the diameter of the wheels, upon which the resistance, which governs the cost of transport and the speed that may be obtained, so materially depends".

The Board accepted his plea for the wider gauge, and then in designing the track he included his second, and less successful, innovation. The longitudinal timbers, forming a continuous support for the rails, were nothing new; they were in fact a perpetuation of the oldest form of railway track dating back to tram roads existing long before the introduction of locomotives. But even in adopting this form of construction Brunel brought some original thinking. Instead of the plain rectangular bars of old, he used a rolled section with flanges extending on either side of the central "rail" portion to spread the load on the supporting timber and provide ease of fixing. Fig. 13, which is reproduced from one of Brunel's own sketch-books (*SB, Miscell.*), shows an interesting refinement in the profile of the head of the rail. While the top surface is flat for $\frac{7}{8}$ inch on the inner edge, there is a short portion tapered at 1 in 10, which provided a valuable centering effect and prevented vehicles from yawing from side to side as they travelled along. Brunel's original sketch shows the rail head solid, but this was subsequently modified into the celebrated "bridge" section, which reduced the weight, without appreciably reducing the carrying capacity, and made the rails easier to roll.

It was, however, not in the cross-sectional shape of the rails that Brunel's principal innovation in track design lay. The arrangement for a double line of rails is shown in another of his original sketches, which was described in an independent report to the GWR Board made in 1838 by Nicholas Wood (MacDermott, 1927):

"Longitudinal timbers of a scantling of from 5 to 7 inches in depth and 12 to 14 inches in breadth, and about 30 feet long are placed along the whole line. Then these timbers are bolted to cross-sleepers or transoms at intervals of every 15 feet; double transoms each 6 inches broad and 9 inches deep being placed at the joinings of each of the longitudinal timbers, and single transoms of the same scantling being placed midway between the joinings. These transoms stretch across, and are bolted to all the four lines of rails. Within the two lines of rails of each track piles of beech are driven from the upper surface of the Railway into the solid ground, so as to retain a firm hold thereof, and the transoms are bolted to the heads of the piles.

"The principle of construction is this: the longitudinal timbers and transoms being firmly held down by the piles, gravel or sand is beaten

Fig. 14. GWR: lamp post

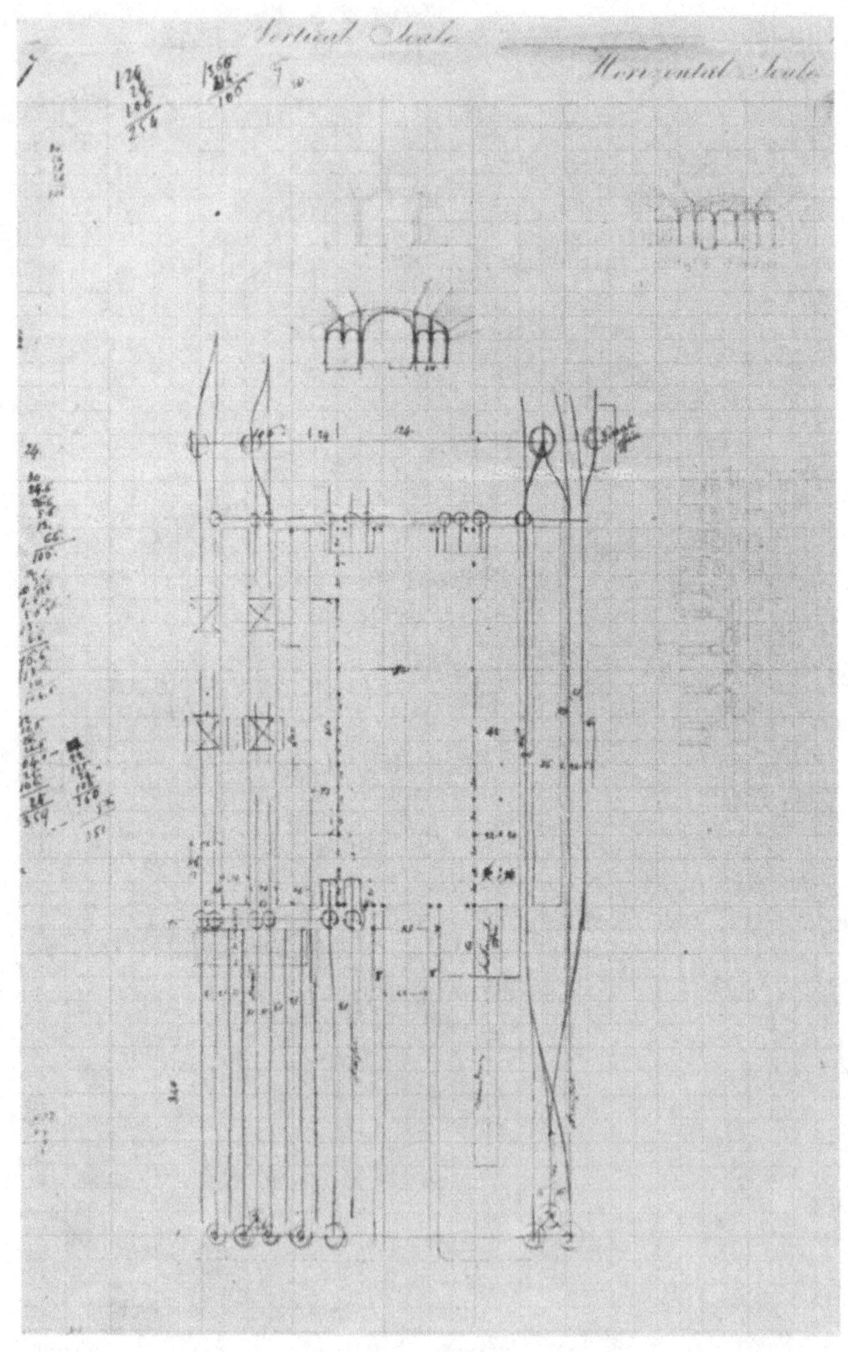

Fig. 15. GWR: Paddington layout

or packed underneath the longitudinal timbers for the purpose of obtaining a considerable vertical strain upon the timbers upwards, and consequently to effect a corresponding firmness of foundation of packing underneath them. Without piles, the longitudinal timbers could not be packed in this manner, as there would be nothing to resist the pressure of the packing except their own weight, and the piles were therefore introduced to hold down the timbers and to render it practicable to introduce a force of packing underneath."

It was an ingenious conception, but it did not work out in practice. The packing was not satisfactory and the longitudinal timbers deflected between successive piles under the weight of a train, and led to a pitching or seesaw motion. The piles were used only on the first section of line opened, between Paddington and Taplow. The connections to them were later severed, and over subsequent stretches of line a heavier rail weighing 62 pounds per yard, instead of the original 43 pounds, was used. This, with improved packing, made the line satisfactory.

Brunel himself designed all the stations, not only in their architectural features, which were often most beautiful, but in their track layouts and detailed facilities for working the traffic. He rapidly became the complete railwayman, planning every single detail of the line. His sketch books are indeed illuminating, in the evidence they provide of the fine detail to which he gave his close attention. He was a consummate draughtsman and there were times when it seemed that he made exquisite little sketches in his books for the sheer love of drawing. How else could one explain the artistic designs for the lamp posts on Bath and Bristol stations contained in his sketch books (Fig. 14)? Among track layouts one finds the plan for the engine house at Bristol and his first essays for the "new" station at Paddington. The original terminus was sited to the west of Bishops Bridge Road, where the present goods station is; what he called the new station is the present terminus which includes many architectural gems of his own design. Fig. 15 shows one of his first proposals for the new station, including many turntables at the inner ends of the platforms to facilitate the traversing of four-wheeled carriages from track to track. There are several different proposals for the roof of the station, all seemingly designed to provide a spectacular vista in the direction of Bristol.

Each of the wayside stations was given individual attention, but as his activities multiplied it was only for the larger ones that he could spare time to make the drawings himself. He seems to have favoured the single platform design, and of this Torquay on the South Devon Railway provides an interesting example (Fig. 16). All the station offices were concentrated in a single block, but nevertheless both running lines were spanned by one of his picturesque all-over roofs. The single platform type of station was evidently tailored to suit the local circumstances; of this the most remarkable example was at Reading, where a maze of crossover tracks developed to provide connection from all the through running lines to the one long platform. While this layout served admirably in the early days of the line, for a town that lay almost entirely on the south side of the railway, its inconvenience from the railway operating viewpoint soon became evident, and in 1853, only twelve years after the opening of the line throughout between London and Bristol, Brunel sketched out *(SB, 33)* two possible proposals for a rearrangement of the tracks and platforms at Reading (Fig. 17).

These sketches are evidently far from complete, but are interesting particularly in his curious treatment of the connections from the "Berks and Hants" line—destined to become, in 1906, the principal main line to the west of England. Presumably to provide easy connections for passengers transferring to or from main line trains, his proposal took a single line between the down and up main line platforms, with the inconvenience of crossing the down main track in the approach to the platforms. The through line for non-stopping trains was carried outside the entire station complex. Both these proposals of 1853 would shock any modern railway operating man by their various intersections of opposing flows of traffic, but they were nothing unusual at that period, and show that in this respect Brunel was no more advanced in his thought than most of his contemporaries.

It is known that in 1848 Brunel himself had in mind an extension of the Berks and Hants line westward from Hungerford, as a shortened and quicker main line to the south-west of England; but any ideas in this direction were scotched by the advance of the narrow gauge London and South Western Railway from Salisbury.

A railway over which great controversy raged was the Oxford and Rugby. It was projected in the aftermath of the "battle of the gauges", which went against Brunel and the Great Western to the extent that no *new* railways were to be permitted on the broad gauge—only extensions of the existing lines. The Great Western was anxious to advance the broad gauge from Oxford to Birmingham and Wolverhampton, but a tremendous fight against this was sustained by

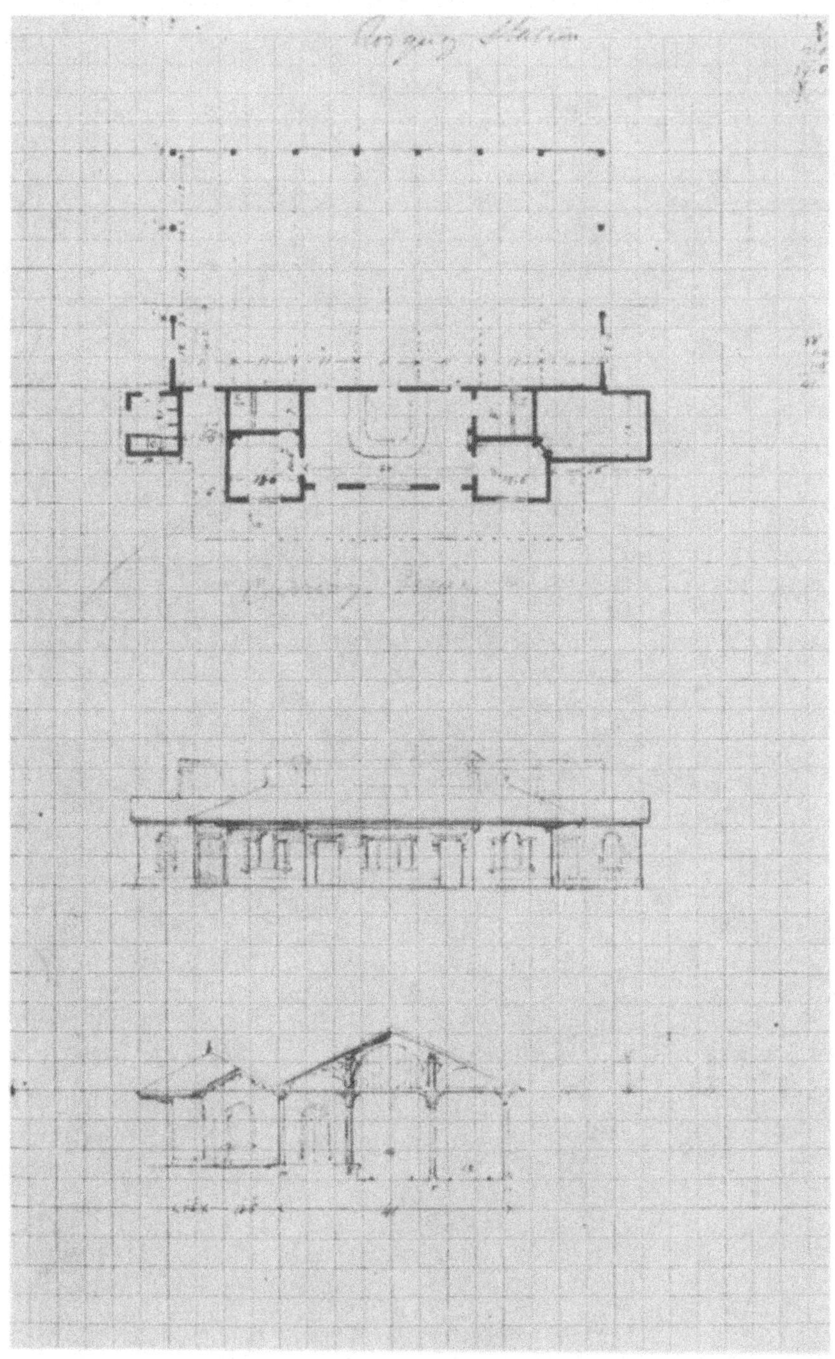

Fig. 16. GWR: Torquay station

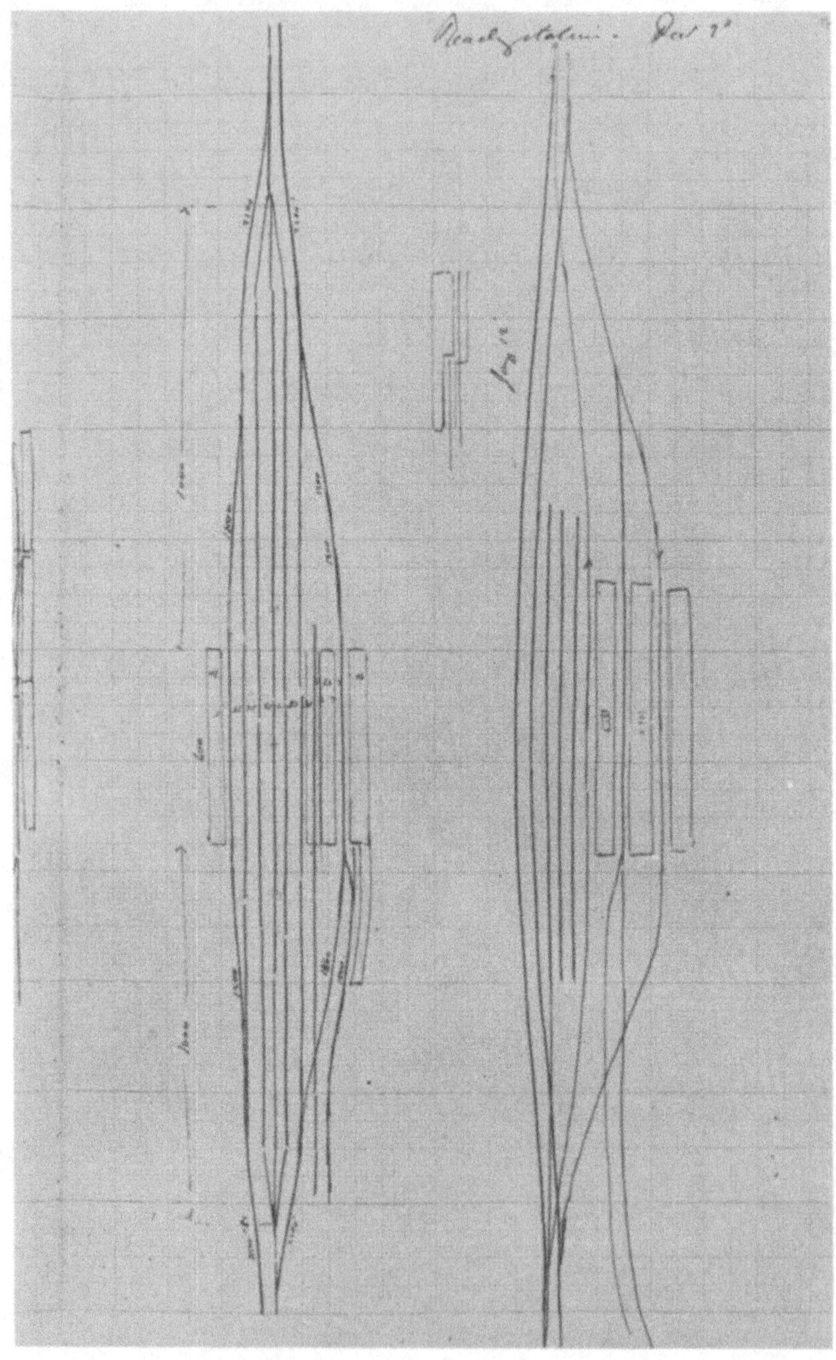

Fig. 17. GWR: Reading layout

the narrow gauge interests, and eventually this line was authorised only on condition that it was laid with both gauges. Brunel's design for the so-called "mixed gauge" track is shown in Fig. 18 *(SB, 25)*.

As the complete railwayman, Brunel at first took personal responsibility for the provision of locomotives, although not to the extent of sketching them out in his notebooks. The only requirement he laid down when inviting proposals from various manufacturers was "a minimum speed of thirty miles an hour". What was actually meant by "minimum" is not clear, but Brunel later wrote:

> "I fixed this velocity so that the engines should be adapted to run 35 miles an hour, and capable of running 40—as the Manchester and Liverpool engines are best calculated for 20 and 25, but capable of running easily at 30 and 35 miles an hour; and fixing also the load which the engine was to be capable of drawing."

There is no doubt also that Brunel unofficially passed on to the prospective builders his own predilection for large wheels; in consequence, while the most usual diameter for the driving wheels of passenger engines on the narrow gauge was 5 feet to 5 feet 6 inches, the first broad gauge engines on the Great Western had wheels of 8 feet and even 10 feet in diameter. Brunel was landed with a collection of freaks. The three engines with 10 foot diameter driving wheels had these made of solid plate, and were not only very heavy but offered a serious resistance to side winds. It was not until he secured Daniel Gooch as locomotive superintendent that Great Western motive power began to assume a reliable and efficient form.

The need for some form of signals to regulate traffic along the line was not at first recognised, and it was thought that the railway policemen could do all that was necessary. At certain locations ball signals were introduced, but were used only for a short time. The vagueness of the working instructions is evident from the following extract from an order of 1840:

> "A signal ball will be seen at the entrance to Reading Station when the line is right for the train to go in. If the ball is not visible the train must not pass it." (MacDermott, 1927)

For many years the absence of any positive indication was part of the signalling code on most British railways, although in the reverse way to the example just quoted in that no signal at all was the all-clear. Brunel, however, holds an honoured place in railway signalling history for having devised an arrangement that gave positive and highly dissimilar indications for all-clear and stop. This was the celebrated "disc and cross-bar" signal, at the top of a tall mast. The "all-clear" was indicated by display of a large red disc broadside on, and "danger" by a lengthy cross-bar at right angles to the disc. The assembly was rotated about a vertical axis, so that the unit not to be seen by the driver was edge on to the line of the railway. When the speed of trains increased, and the need to provide some form of advance warning of a danger signal ahead became urgent, Brunel was at first not so successful. He devised a curious form of "flag" signal, with red and green curtains attached to rings travelling on an iron bow frame, and worked by cords and pulleys. Although rather expendable they worked reasonably well in calm weather, but were torn to ribbons on windy days. To obviate this trouble he substituted the "fantail" board—another signalling speciality that was entirely his own.

As the broad gauge system was extended south-westwards from Bristol, certain curiosities in Brunel's constructional practice became accentuated. To admit of continuous high-speed running he had laid out the original line from London to Bristol on very easy gradients, except for two relatively short sections where the inclination was 1 in 100, namely the Dauntsey bank beginning shortly after the 85th milepost and through the Box Tunnel. To what extent he could have avoided these sharp inclines by deviations in the line is conjectural, but there was an element of inconsistency in his route planning that Dauntsey and Box Tunnel do not entirely explain. To get an almost gradeless line over the first 85 miles he chose to take the Thames valley from Reading, and then the Vale of the White Horse, bypassing the country towns of Wallingford, Abingdon and Wantage *en route*. The more obvious course, to tap sources of intermediate traffic, would surely have been to take the Kennet valley from Reading and pass through Newbury, Hungerford, Devizes and Bradford-on-Avon on the way to Bath. The gradients would not have been difficult, the 1 in 100 inclines at the western end would have been avoided and the curves in the Avon valley between Bradford-on-Avon and Bathampton would not have been a serious hindrance at the speeds run. On the Bristol and Exeter line he made a frontal attack on the Blackdown Hills, with a very stiff pull at 1 in 80 up to the Whiteball Tunnel, where no more than a moderate deviation and a longer tunnel at the summit would have provided a line much easier to work.

In approaching Exeter one comes to what is perhaps the most inexplicable of all his ventures into railway engineering: the atmospheric system of traction. It arose from his inspection of an experimental length of 1½ miles laid on a piece of waste ground in west

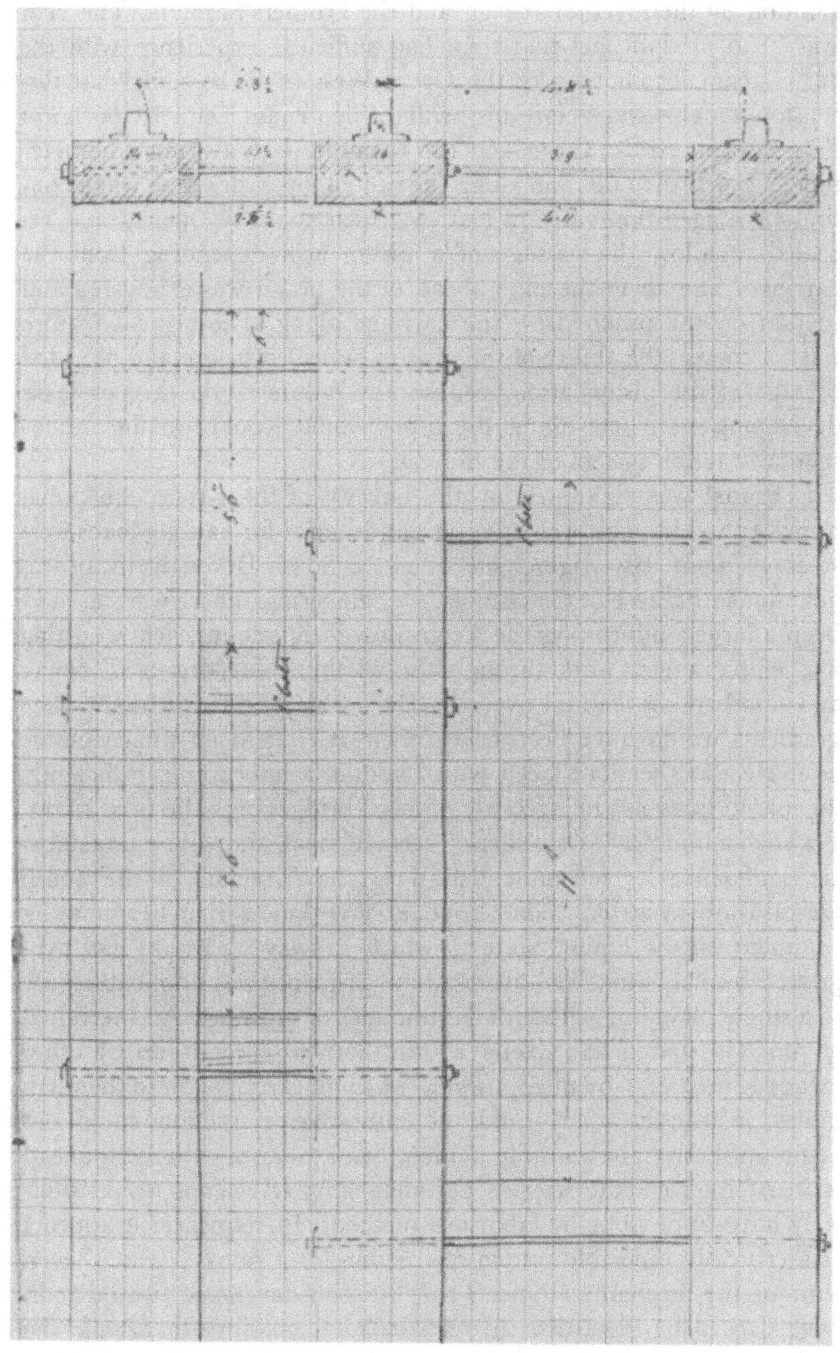

Fig. 18. GWR: mixed gauge

London by the inventors Clegg and the brothers Samuda. The year was 1840, and Brunel had then had sufficient experience with the early steam locomotives of the Great Western to be somewhat disillusioned. This was before his brilliant lieutenant Daniel Gooch got fully into his stride. Clegg and the Samudas laid a 9 inch diameter pipe between the rails, and along the slotted top of this pipe was a continuous hinged flap valve of leather which could be opened and resealed to allow the passage of a piston arm depending from the carriage. The air in the pipe ahead of the piston was exhausted and suction on the piston drew the carriage along. The critical feature was, of course, the ability of the pipe to be sealed before and after the passage of the piston arm, because any failure to do this, or leaks developing elsewhere along the pipe, would jeopardise the fundamental traction system of the railway.

Brunel was fascinated by the novelty of the system, but what appealed to him most was its apparent capacity for hauling loads up a steep gradient. His original survey of the South Devon Railway was abandoned because of the cost of the engineering. There were to have been a large viaduct over the Teign near Teignmouth, heavy cutting and embankment work through the northern outskirts of Torquay and another big viaduct over the Dart valley near Dittisham. The gradients would have been relatively easy, but at an unacceptable cost. He was therefore faced with building a hilly route, sufficiently far inland to avoid the necessity of large bridges over the tidal rivers. For such a line the "atmospheric" seemed ideal, and with characteristic confidence he recommended it to the Directors of the South Devon Railway in 1845. That he could have done so is an illuminating comment on the limited state to which railway operation had progressed by that time. The atmospheric principle was applicable only to a single plain line without any junctions or crossings. At a terminal or junction station like Exeter a train had to be made up by other means of traction—by men pushing the small carriages on to the turntables or transfer roads—and the atmospheric traction could not begin until the train was fully made up and the tube exhausted ahead of it. At intermediate stations any detaching of wagons would similarly have to be done by laborious methods. In retrospect it appears as a critically inflexible system of traction for a railway which, even without the mechanical defects that so soon developed, would soon have had to be discarded through its total inability to handle the traffic.

Brunel entered on the construction with his usual vigour and enthusiasm. The pumping stations for creating the necessary vacuum in the tube were situated about three miles apart and each was a little cultural masterpiece of a building; and in carrying the line round the coast beneath the picturesque red sandstone cliffs of Dawlish and Teignmouth he had, for the first time in his railway building, to contend with the sea. On the exposed stretches he built substantial stone walls; but, while these were not high enough to prevent waves breaking over passing trains in stormy weather, the principal difficulties arose from the instability of the cliffs themselves, and on occasions the line has been blocked by major landslips. The entire section from Exeter to Newton Abbot was single tracked, although being laid to the broad gauge the numerous tunnels between Dawlish and Teignmouth did not need very much enlarging when the time came to double the line after the conversion to standard gauge.

The atmospheric system of traction failed through rapid deterioration of the leather flap valve along the top of the tube. The system lasted for less than a year on the South Devon line: from September 1847 to June 1848. At the outset its performance was promising, and speeds of more than 60 miles per hour are fully authenticated. But in the winter the leather sometimes froze, rats gnawed it and the sea air completed the destruction. Brunel was a big enough man to admit defeat and to bear the considerable financial loss that his personal investment in it involved. Really, however, it was fortunate that the technical failure came so soon because, even if mechanically perfect, one cannot conceive how the steadily increasing traffic over the line could have been operated on such a system of traction. It is regrettably easy to be wise more than 100 years after the event, but it does seem that over the South Devon Railway Brunel's remarkable perception deserted him—albeit briefly. Its inflexibility at junction points, and the need to have other forms of motive power for shunting and remarshalling would inevitably have involved much delay, however fast the trains ran intermediately. The long stretches of single tracked line would in any case have severely restricted its traffic-carrying capacity. It was not as though Brunel was driving his railway out into a virgin, undeveloped countryside. Ahead lay the major seaports of Plymouth and Falmouth, and the potential traffic from the Cornish tin mines.

It was the potentialities of the "atmospheric" for surmounting steep gradients that lured Brunel into laying out the most extraordinary stretch of line to be found on any first class main route in Great Britain: the 6½ miles between Newton Abbot and Totnes. The countryside is certainly hilly, but there the man who had driven his original main line straight through such an obstacle as Box Hill followed the lie of the land, not only on exceptional gradients up to a maximum of 1 in 36, but also on sŭch a succession of sharp curves that maximum speeds downhill have had to be restricted to 40 miles per hour. The pipe for the atmospheric was laid as far as Totnes, but there is no record of any train service having been attempted west of Newton Abbot. A contemporary record of what the atmospheric line looked like exists in a comprehensive series of water colour drawings, covering nearly every mile between Exeter and Totnes, in the library of the Institution of Civil Engineers.

Before he had become involved with the atmospheric system on the South Devon Railways he was cross-examined by the Gauge Commissioners (Royal Commission on Railway Gauges, 1845) as to why he had recommended gauges of less than 7 feet for other railways of which he was the Engineer. One of these was the Taff Vale, which was laid to 4 feet 8½ inches. He was asked why, and replied:

> "One of the reasons, I remember, was one which would not influence me now; but at that time I certainly assumed that the effect of curves was such, that the radius of the curve might be measured in units of the gauge, in which I have since found myself to be mistaken. Then I expected to have to lay out the line with a succession of curves of small radius, which is the case as the line is laid out; and I assumed that the narrow gauge was better than the wide gauge as regards curves. I do not remember whether connexion with any other railways there, or likely to be there, influenced me."

Certainly he seemed to have overcome any inhibitions over curves when he came to lay out the South Devon Railway, beside which the Taff Vale could be regarded as a straight line!

At the time of the Gauge Commission in 1845 Brunel was Engineer to the Genoa–Turin line being built by the Sardinian Government, for which he recommended the 4 foot 8½ inch gauge. He told the Commissioners:

> "The reason that led me to adopt it was this, that I did not think that either the quantities or speeds likely to be demanded for many years to come, in that country, required the same principle to be carried out that I thought was required here: and I thought it very important that they should secure the goodwill of certain other interests which would lead into and out of this railway; and as a question of policy as much, as of engineering, I advised them to adopt that gauge. I thought it wise to conciliate the interest of the Milan and Venice Railway and others which are likely to be connected with us."

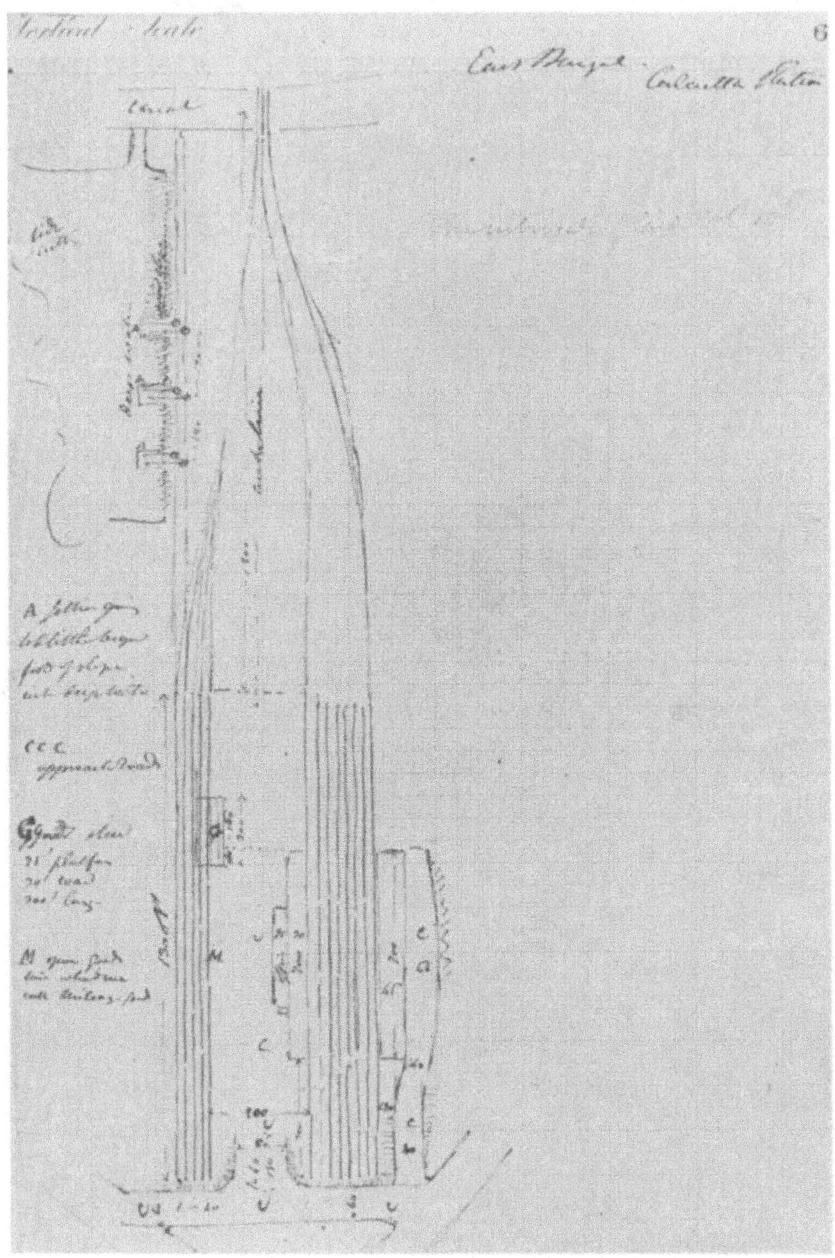

Fig. 19. East Bengal Railway: Calcutta station

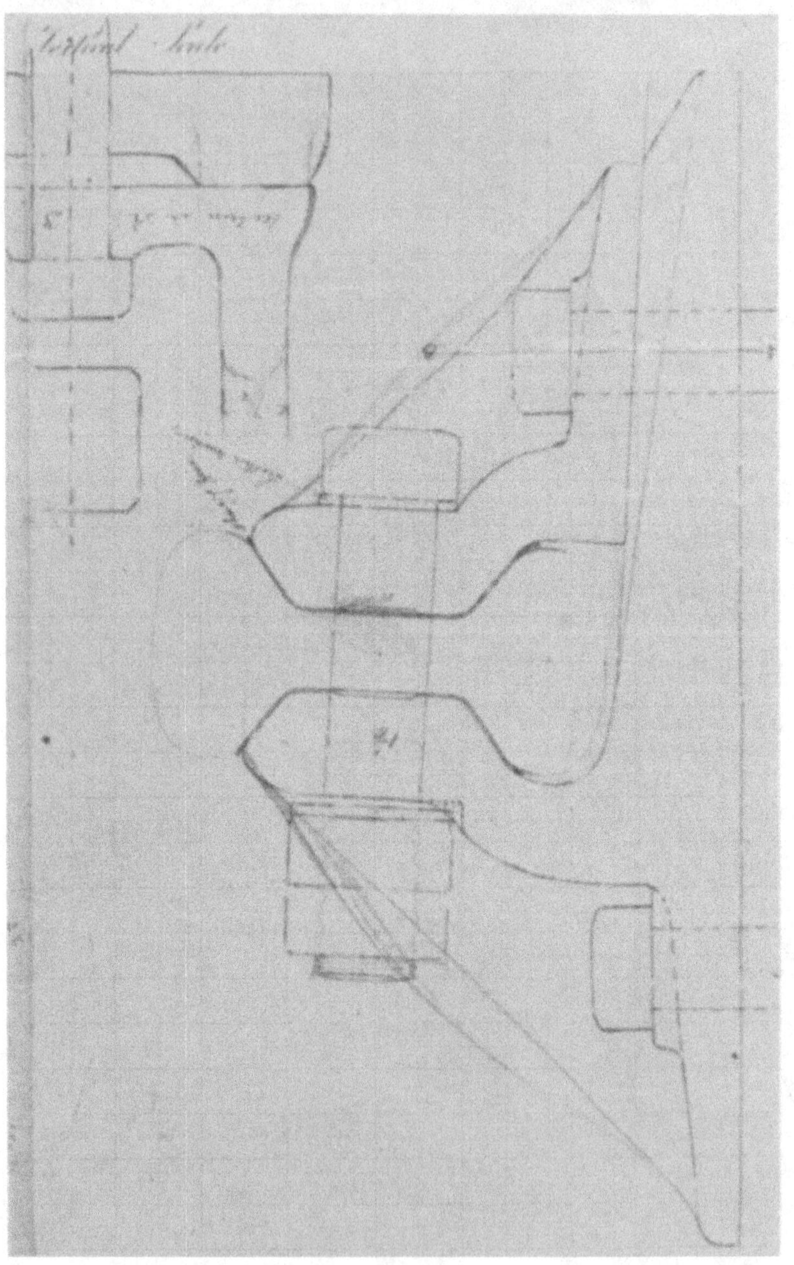

Fig. 20. East Bengal Railway: rail chair

There had been nothing in the way of conciliation towards the interests of other railways in Great Britain. There is little doubt that he was so convinced of the superiority of the broad gauge that he imagined that, once people saw it in operation, conversion of all existing lines would only be a matter of time. On the 4 foot 8½ inch Genoa–Turin line, however, he had some gradients that make even those of the South Devon Railway pale into insignificance. Inland from Genoa there is a mountain range parallel to the coast, and this had to be surmounted in the first 40 miles of the railway to the north. The original double tracked line, which was later duplicated, has a maximum gradient of 1 in 27 and passes under the Ligurian Apennine range in the Giovi Tunnel, 2 miles long. It was this route that was engineered by Brunel. Although at the time no great development of the traffic was foreseen, the development when it did come was such as to require the provision of two additional running lines, and to lessen the difficulties of haulage in such a terrain the newer line was built on easier gradients that involved a tunnel 5 miles long under the mountain range.

At the time of his death in 1859 Brunel was engaged in designs for the Eastern Bengal Railway, and Fig. 19, reproduced from his sketch-book *(SB, 13)*, shows his proposals for the terminus of the line in the Sealdah district of Calcutta. But a point of outstanding interest on this work is the design of the rails, dated 1858. His great friend and professional rival Joseph Locke had invented the double headed type of rail, so conceived that when the upper surface was worn it could be taken out and turned over so that what had hitherto been the underside could be used. By this it was hoped to secure roughly double the life of an ordinary rail. But in resting on its "chair" Locke's rail sustained abrasions on its under surface through the transmission of loads in the passage of traffic, and when turned over the surface that had been underneath was unusable.

Brunel took the double headed principle propounded by Locke, and after applying it first on the South Wales Mineral Railway designed a special chair and clamp (Fig. 20) for use on the Eastern Bengal. The important feature was that the under surface of the rail did not rest on the chair at all. The chair was shaped to the profile of the rail, and the downward load was taken on this and the clamp, which was bolted up tight. Sheets of zinc were inserted between the rail and the clamps, and the bolts were pulled up against leather washers—a most ingenious design that would preserve the under surface of the rail until the time came for turning over. It has unfortunately not been possible to ascertain the mileage actually laid in India with this type of rail. Brunel never visited the works during the time of construction, and he died before the railway was opened to traffic.

V

Arch Bridges

PROFESSOR J. B. B. OWEN

Isambard Brunel (1870) devotes chapter seven of his biography of Isambard Kingdom Brunel to his father's railway bridges and viaducts. He deals first with those bridges built of brick and masonry—the arch bridges which form the subject of this chapter. The first of these which he mentions is that over the River Brent at Hanwell (Fig. 21), later called the Wharncliffe Viaduct. He describes it as "the first of Mr Brunel's important railway works . . . a handsome brickwork structure" (as indeed it still is, particularly when viewed, over green fields, from the main road just west of the Viaduct Inn, Hanwell) "65 feet high, with eight semi-elliptical arches, each 70 feet span and 17 feet 6 inches rise". Brunel's general calculation book *(CB, 1837)*, which forms part of the manuscript collection of Bristol University, also contains under the heading "Brent Viaduct" several pages of drawings and calculations relating to these arches. These calculations are discussed later.

But to provide sustained news interest bridges must fail or be considered to be in a critical state. It is then not surprising to find that all Brunel's biographers deal in some detail with his Maidenhead bridge. Most gloss over the troubles which occurred with the eastern arch of this bridge, but Fowler (Mackay, 1900), when a young engineer, considered them most serious. Later he was responsible for retaining Brunel's arch shape when the bridge was widened. The calculation book *(CB, 1837)* also contains drawings and calculations for this bridge and, going westward along the main line, for Thorney Broad Bridge, which is over the River Colne just above Thorney Weir, and for the central quasi-Gothic arch over the Avon just below the weir above Temple Meads railway station at Bristol. The drawing

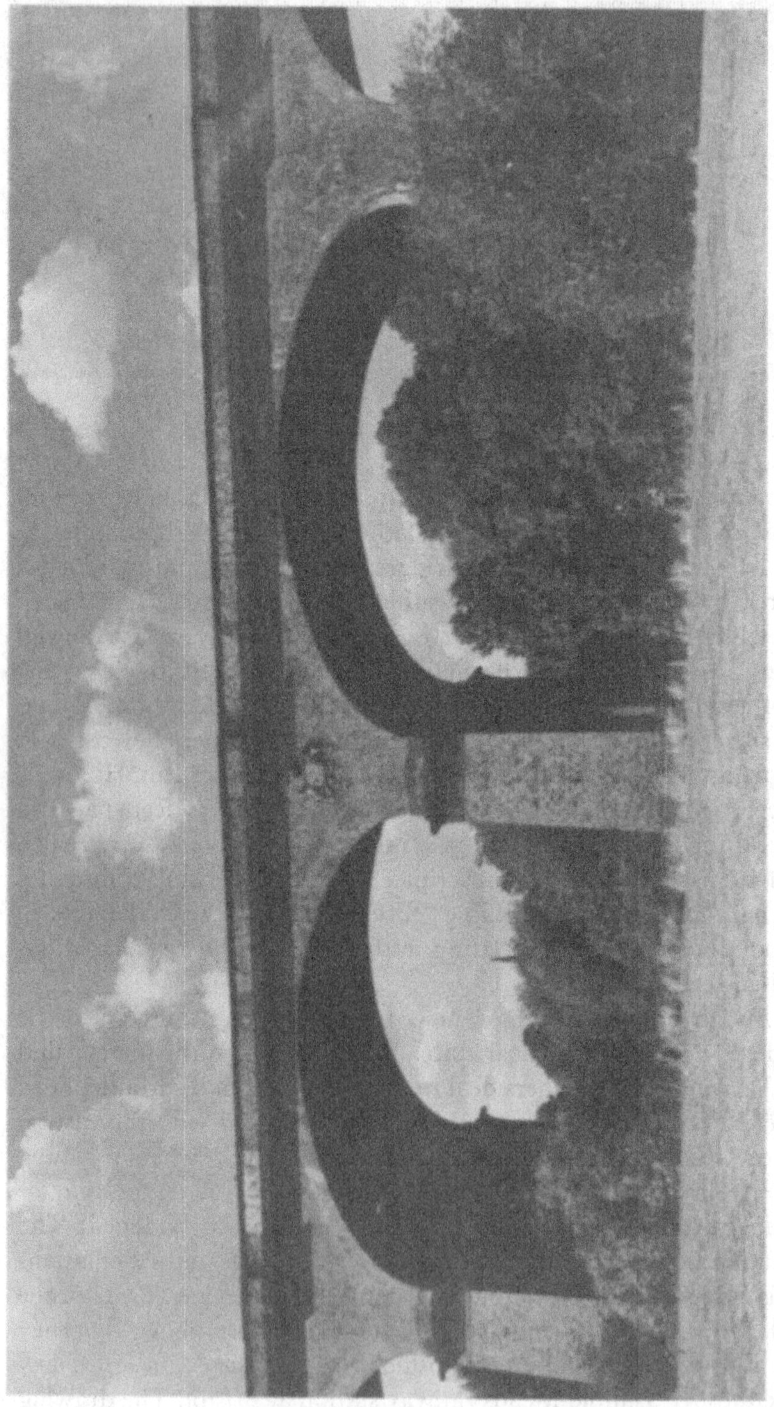

T. L. J. Bentley

Fig. 21. Wharncliffe Viaduct at Hanwell

in the book, however, is of a segmental arch, which forms the central span of this bridge. Its elegant Gothic quoins, which are now almost hidden by latticed girders on both sides, complete the transformation of the central segmental arch into the Gothic arch of Fig. 22 (sketch-book, *SB, 1835, 5 29/5 35*; see also Bourne, 1846) which is one of the better of the many sketches of arches which abound in the Bristol manuscripts. The emery-paper and sandpaper stuck inside the covers of the bound volumes of the manuscripts provide some evidence of Brunel's workmanlike approach to his draughtsmanship. What a pity it is that the artistic flair so evident in the drawings of a two-span skew Gothic arch bridge (Brees, 1837, plate XLII) over the Floating Harbour on the approach to Temple Meads railway station, is now almost completely hidden by segmental arches which make the bridge wider on both sides.

I. Brunel (1870) indicates that the crown of a 120 foot span, 20 foot rise segmental arch bridge "over the New Cut at Bristol . . . sunk much more than expected",[32] but says nothing about the replacement of an arch bridge (the Somerset Bridge) over the Parrett which MacDermott (1927) describes. The drawing store of British Rail at Paddington has a drawing of this bridge, part of a priceless collection of Brunel's drawings which are at present housed there. A brief description of some of Brunel's arch bridges can also be found in Brees (1837), Simms (1838), Bourne (1846), Sekon (1895), MacDermott (1927) and Rolt (1957).

The problems with the Maidenhead bridge occurred in the spring of 1838. Considerable numbers of arch bridges were built in the mid nineteenth century but it was not until 1846 that W. H. Barlow, in a paper to the Institution of Civil Engineers, indicated the nature of the mechanisms which form when arches fail, and displayed models illustrating his arguments. These ideas were developed a little further by Snell (1846) and in the middle of the present century again by Pippard and Baker (1962) and others eventually to yield modern ideas of structural collapse. Studies by Ruddock (1974) suggest that failures of arches by rising at the crown are historically not uncommon, but Heyman (1969) considers that a design has indeed to be clumsy to fail in this way! In this chapter what can be gleaned of the nature of the Maidenhead and Somerset bridge troubles is outlined and the ideas underlying the calculations in the Brunel manuscripts are examined. It would appear that Brunel might have developed an almost direct and safe approach to arch design, which would nowadays be described as "statically admissible". Brunel's practice of using what Fowler describes as "very excellent and scientific centering" avoided disaster in their construction (Mackay, 1900).

According to MacDermott (1927), where the Maidenhead bridge was constructed the Thames was

> ". . . not quite 100 yards wide between low banks with a small shoal in the middle. Brunel designed a brick bridge of two of the flattest and largest arches that had ever been constructed in brickwork. Each arch had a span of 128 feet with a rise of only 24¼ feet, the centre pier being placed on the shoal. Four semi-circular arches at each end to provide for floods, one of 21 and three of 28 feet span, completed the structure. Work was begun early in 1837 by the contractor, one Chadwick, and before the end of the year the arches had been turned. In February 1838 the Directors reported that the bridge was in a very forward state and the centering would shortly be eased. It was eased in the course of the spring, and to the delight of the critics, who had all along declared that such flat arches could not possibly stand, the eastern arch soon showed signs of distortion. The western arch, however, annoyed them by standing perfect."

Mackay (1900) quoting from a letter of Fowler, dated 11th May, 1838, paints a gloomy picture of these "signs of distortion". Fowler writes:

> "The bridge I went to see is of a novel construction, being built of brickwork set in cement for two elliptical arches of 128 feet each, and is now, I am sorry to say, in a dangerous situation. The centering has been slackened and the arches have followed it for 5 inches at the crown, and now rest on them, but not as you would suppose at the crown of the arch, but at about 15 feet on each side, while the crown is 2½ inches clear of the centering. This settling has made the outline of the arch perfectly Gothic and totally destroyed its symmetry, but the worst of the matter is that the arch which has been turned in half-brick rings has separated at every other joint, and consequently the arch is further there weakened by the lower part of it leaving the upper part. The outside of the arch shows eight half-brick curves, but the inside has thirteen. The spandrel wall of one arch has cracked from top to bottom, so likewise have all the walls for backing up the inside of the arch, both occasioned of course, by the settling. The foundations are on rock, and therefore very unyielding. The centering is a very excellent and scientific one, the whole of the timbers being in a state of thrust. What will or can be done is too much for me to say."

There is no evidence, however, of panic action by Brunel. The defective arch was resting on its "very excellent and scientific centering". Earth was being taken over from the Buckinghamshire side by

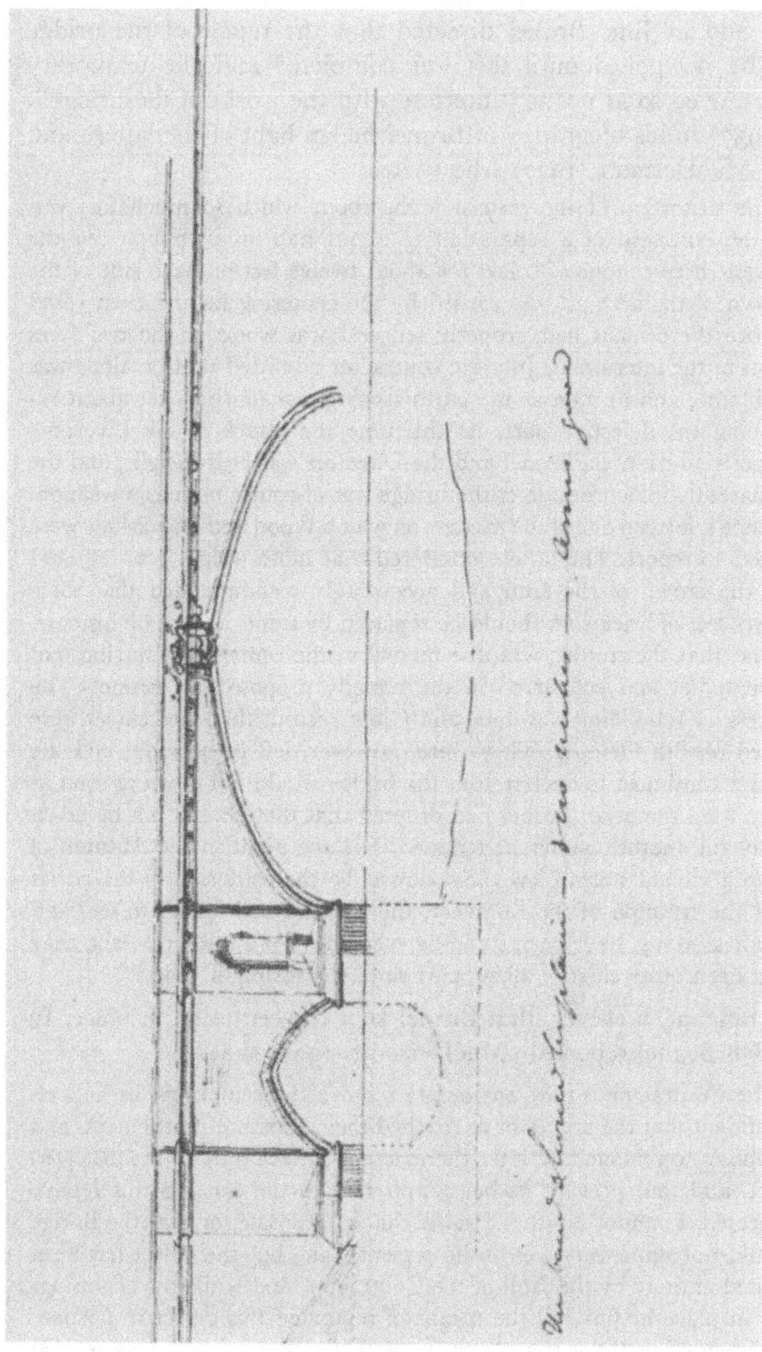

Fig. 22. Arch near Temple Meads station

temporary roads to form the embankment immediately west of the bridge, and in June Brunel directed that the repair of the bridge should be postponed until this was completed and the temporary roads removed so as not to "interfere with the works of the bridge". Rolt (1957) in his biography of Brunel makes light of the failure and so does MacDermott (1927) who writes:

> "The distortion of the eastern arch, about which so much fuss was made, consisted of a separation of about half an inch between the lowest three courses of bricks for about twelve feet on each side of the crown of the arch. It was caused by the centering having been eased before the cement had properly set, and was worse at the two faces than in the interior. In July the contractor admitted that he alone was to blame, and as soon as the earth tipping was finished, set about replacing the defective part. At this time the attack of the Liverpool opposition party on Brunel and the Directors was at its height, and the apparently insecure state of the bridge was of course used as a weapon. Hence it formed one of the matters on which Wood and Hawkshaw were asked to report. The latter considered that more weight was required on the crown of the arch and accordingly recommended that some thirty feet of brickwork should be replaced by stone. Wood, on his part, found that the trouble was due merely to the cement not having had time to set and concurred in the remedy proposed by Brunel. The necessary rebuilding was done and Gibbs records that the centres were eased on 8th October. They were, however, left in position, and the critics continued to declare that the bridge would fall down as soon as they were removed. Brunel had ordered that they should not be taken away till another winter had passed, but one night in the Autumn of 1839 a violent storm blew them down. To the confusion of the critics and the triumph of the Engineer, the arches stood for all to see, and needless to say, have been standing ever since. As a matter of fact, they had been quite clear of all support since the spring of 1839."

It is significant, however, that Brunel kept the centering in place. In June 1838 Brunel reported (MacDermott, 1927) that:

> "The Contractor is now anxious to remove the centerings, as he feels confident that the arches have reached their maximum settlement, and proposes to remove and alter the external surface only of the distorted part, and thus prevent its being apparent to the eye. To this I have objected. I cannot say that I think the work unsafe, or that the Bridge would not stand very well in its present state, but the defect has been caused entirely by the fault of the Contractor, and while the centerings are in place he has still the means of removing this defective portion, and rendering the eastern arch as perfect as the western now is, and such as he contracted to make it. I propose therefore to direct Mr

> Chadwick to tighten up the centres of both arches and to remove and replace so much of the eastern arch as upon close examination during the progress of this work I may consider defective . . ."

T. M. Smith (1846) in the spring of 1838 (the time when the centering on the Maidenhead bridge was eased) presented a paper to the Institution of Civil Engineers in which he describes the failure, some 85 years earlier, of a William Edward's 140 foot single-span masonry arch bridge, over the Taff at Pontypridd, which probably had a rise of 35 feet:

> "From the great rise of the arch, it was necessary to raise the roadway at each end, in order to facilitate the access. The quantity of material in the centre of the arch was so small, in proportion to that laid over the spandrils, that the pressure upon the haunches forced up the crown, and the whole was again reduced to ruins."

Perhaps this upward movement of the crown may have prompted Hawkshaw's recommendation that "more weight was required on the crown" of the Maidenhead bridge (MacDermott, 1927). It seems that Edward added such weight in his successful single-span arch over the Taff at Pontypridd and also introduced three cylindrical holes to lighten the spandrels, and in his bridge there is some evidence (Owen, 1939) that this may well have been necessary. Although not externally apparent in the Maidenhead bridge, Brunel's design also took the weight off the haunches. Drawings available in the British Rail drawing store at Paddington indicate that longitudinal walls were used "as ribs" in the spandrels which are not solid. These ribs are presumably the "walls for backing up the inside of the arch" which Fowler indicates were "likewise" cracked from "top to bottom" (Mackay, 1900). It is significant that although the use of these longitudinal ribs should have increased the effective depth of the arch ring considerably, they did not prevent failure and they thus provide an exception to Ruddock's (1974) conclusion that "the parallel wall method of construction had solved the problem of instability of arches".

By 1846 the Maidenhead bridge had been accepted, and Brunel receives honourable mention by Sir John Rennie in his presidential address to the Institution of Civil Engineers:

> "Brick has been much used for bridges . . . and, latterly, it has been carried to a far greater extent by Brunel in his bridge across the Thames at Maidenhead, for the Great Western Railway . . . they are built wholly of brick and Roman cement."

Sir John also remarked in his address that:

> "Additional strength has been given to brick structures, by the intro-

> duction of bands of thin hoop iron between the courses; this improvement was first generally introduced by Sir M. I. Brunel."

The descriptions of the Maidenhead failure by Fowler and others suggest, however, that such bands were not used in the Maidenhead bridge. Examination of the under side of the Brunel arches at Maidenhead indicates alternate header and stretcher rings, and would suggest a construction of sensibly successive layers of rings of the arch of the kind which would permit the arch ring to separate as described by Fowler (MacDermott, 1927). It may be, however, that the mystery surrounding the Maidenhead bridge failure in 1838 eventually set in motion the investigations of W. H. Barlow and Snell which were published in 1846 and which really dealt with what would nowadays be looked on as the collapse of arches by their becoming mechanisms.

Barlow early on in his paper states (correctly) that the essential element of equilibrium, when the only force acting is gravity, is "that the horizontal forces in any part of the curve (or the arch) are equal to each other". He states that this had not been pointed out previously although it was probably known. Indeed it is a fundamental concept used by Brunel in his manuscripts of 1837. Commenting on Barlow's paper, Brunel remarks that:

> "In a very large arch, with a small rise, the line of pressure must be confined within very narrow limits."

He should have emphasised, as Barlow pointed out, that the thickness of the ring rather than the rise governs the accuracy required, but in emphasising the need for accuracy he made an important point which explains in part why, in his manuscripts of 1837, he eventually uses numerical methods rather than the less accurate methods of graphic statics of his contemporaries.

The calculations in his general calculation book (*CB, 1837*) give only an occasional comment on the procedure used, but from what is done the underlying principles and assumptions can be largely inferred. In strength calculations for arch bridges the arch ring proper is usually assumed to be the only part providing its strength. This is a safe assumption not too far from the truth if the centering is struck before the mortar has set. Some assumptions have also to be made on the nature of the loading on the arch. Brunel initially assumes for the Brent[33] and Maidenhead bridges that the vertical loading on the arch is the weight of elements of the bridge obtained by dividing the arch "at right angles to the line of pressure" or at "the angle of" an arch "joint" (which might not be precisely the same thing) and then vertically through the spandrel, as in Fig. 23, which although for

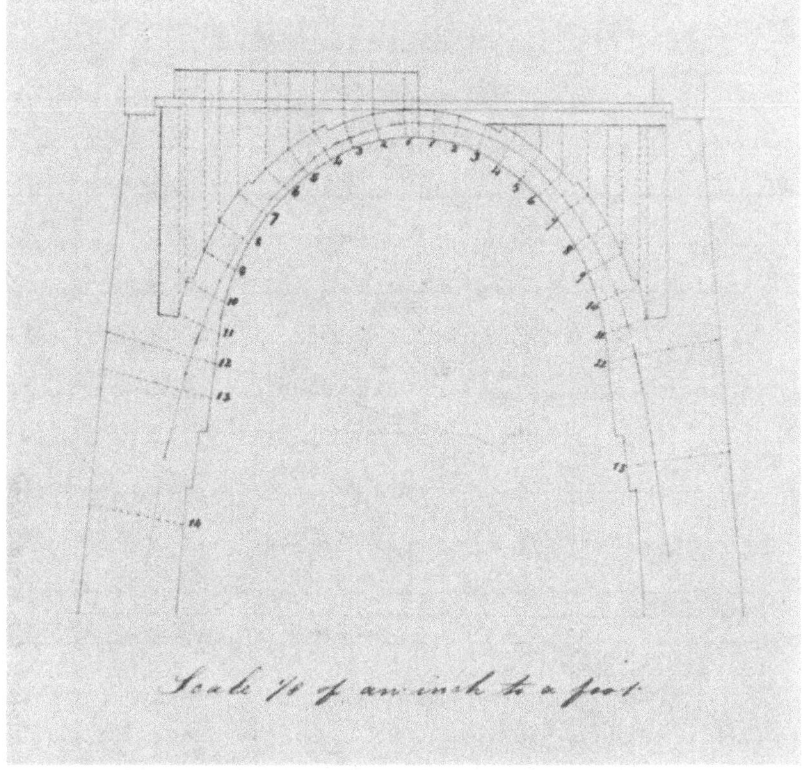

Fig. 23. Arch analysis for Hanwell

only one span may well be a drawing on which were based the six arches in the approach to the Wharncliffe Viaduct. An assumed "angle" at a particular point of the arch ring enables Brunel to calculate by statics the horizontal thrust in the arch. The "thrust line" in, i.e. the funicular polygon for, the arch can now be drawn starting with a horizontal at the crown and diverting this downwards, where it meets the line of the vertical load from one of the elements into which the bridge has been divided. This is clearly seen in the left hand element between 6 and 7. For static equilibrium to be obtained, the point where the polygon changes direction in this element should be the centre of gravity of this element and this would hardly seem to be the case here. The point where the polygon changes direction is critical in determining its shape. The polygon on the left hand side of Fig. 23 comes very close to the intrados at points 6, 7 and 8 and Brunel may not have been satisfied with the result, for he proceeds to calculate what the load distribution should be to make the thrust line

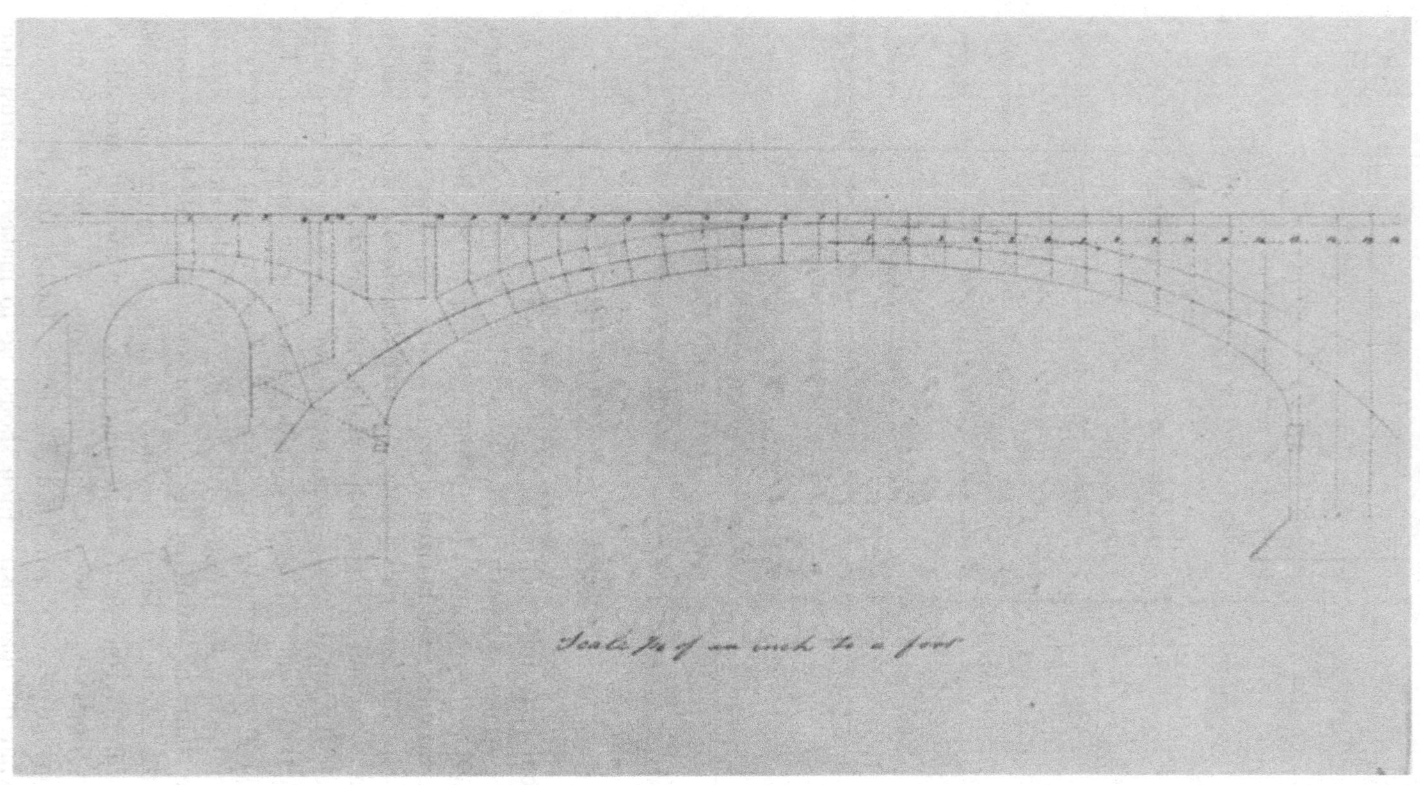

Fig. 24. Arch analysis for Maidenhead

coincide with the centre line of the arch. Still not satisfied perhaps, he starts again and divides the arch into elements by vertical lines running through the whole depth of the bridge (as on the right hand side of Fig. 24 for the Maidenhead bridge). The weights of these elements enable him, without making any further assumption, to calculate "the weights of the half arches", i.e. the shear on vertical sections of the bridge.

At this stage a most novel feature is introduced which avoids the necessity of finding the centre of gravity of each of the elements into which the bridge is divided. The shear is represented by a polynomial of the form

$$W = Ay + By^2 + Cy^3 + Dy^4$$

where y is measured horizontally from the centre line of the arch and A, B, C and D are constants chosen so that the polynomial is correct at four points on the span. Brunel checks that the error at other points is negligible. For the Brent viaduct (short span arches near Hanwell station) he quotes the result

$$W = 42y - y^2/7 + y^3/6$$

This weight has to be resisted by the vertical component of the arch thrust, the horizontal component of which is H. Simple statics then gives the result that

$$H\,\mathrm{d}x/\mathrm{d}y = 42y - y^2/7 + y^3/6$$

and integration gives

$$Hx = 21y^2 - y^3/21 + y^4/24 + C$$

where C is constant. Brunel now chooses to place the thrust at the mid depth of the arch at the centre of the span, where x and y are zero. This makes C zero. He also decides that the thrust will pass through the mid arch ring depth at about quarter span, where the earlier calculations indicated that it approached the intrados. Substitution of the values of x and y for this point then give $H = 441$ and so the equation

$$441x = 21y^2 - y^3/21 + y^4/24$$

from which the shape of the thrust line can be immediately drawn. In important parts of the bridge he thus establishes that it is possible to have a thrust line which is very close to the centre of the arch ring. A modern arch designer might well take such a calculated value for the horizontal thrust in the arch and adjust flat jacks in the abutment until this value was obtained. Brunel effectively demonstrates that if this thrust could be obtained then the arch would be safe. The horizontal component of the thrust in the arch represented by the number 441 he notes is equivalent to a "pillar 24½ feet high".

At this stage he turns to the Maidenhead bridge (Fig. 25) (and the calculations again seem to have been "copied for Brodie, 29 April 1851"). The first calculations with the thrust line made normal to the "voussoirs" (left hand side of Fig. 24) give the horizontal thrust as equivalent to a pillar height of 213 feet and a thrust line which is very close to the centre line of the arch ring. This is followed by the statically more acceptable polynomial approach, already described, in which the thrust line is made to pass through mid arch ring depth at the centre line and a point close to the end of the span (right hand side of Fig. 24). The horizontal thrust he obtains is the equivalent of a pillar of 205 feet (10.2 tons per square foot)—a very low stress compared with the probable crushing strength of the bricks which could be expected to be well over 60 tons per square foot. The thrust line (Fig. 24) is again everywhere near the centre line of the arch ring. There is nothing here except the intensity of the pressure (which is almost ten times that obtained for the Brent viaduct but still low in relation to the strength of the bricks) which would suggest that trouble would arise.

These results indicate why when trouble did arise Brunel did not attempt to load the crown of the Maidenhead arch as suggested by Hawkshaw and why, with good centering, he chose to wait for the weight of the bridge to consolidate the abutments (in case of any movement there) and for the mortar to set. There would be no trouble if he could but obtain his calculated horizontal thrust from the consolidated arch and abutments.

Brunel's manuscript calculations on the Maidenhead bridge stop here. He proceeds next, using only the more acceptable polynomial approach, to check Thorney Broad (88 foot equivalent pillar), the large spans of the Brent, i.e. Wharncliffe Viaduct proper (84 foot equivalent pillar) and the Avon bridge (156 foot equivalent pillar); the thrust lines for the latter again run close to the centre line of the voussoirs.

There are no calculations in the manuscripts relating to the Somerset bridge over the Parrett near Bridgwater; this bridge also gave Brunel some trouble. It was:

> "a masonry arch of 100 ft. span with a rise of only 12 ft., nearly twice as flat as the much criticised brick arches at Maidenhead. This was begun in 1838 and finished in three years" (MacDermott, 1927).

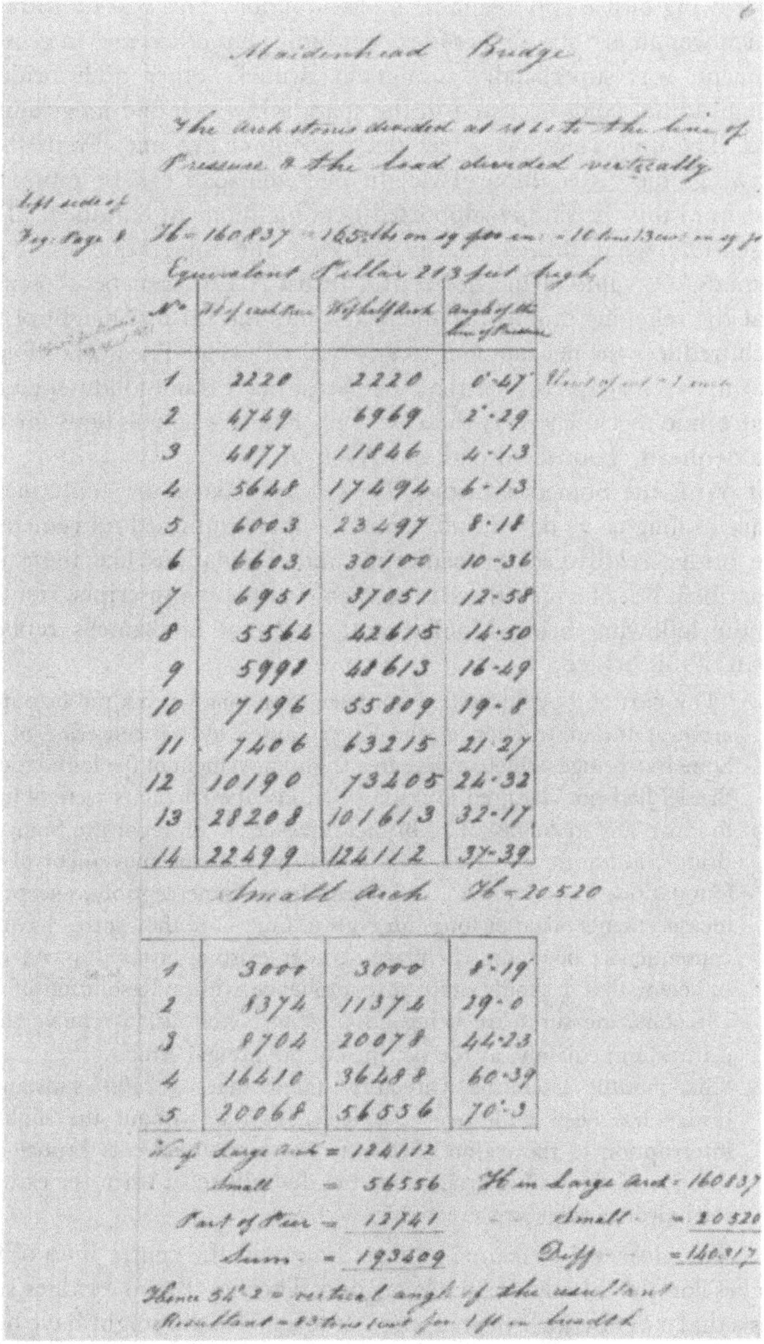

Maidenhead Bridge

The Arch stones divided at rt ∠s to the line of Pressure & the load divided vertically

left side of Fig. Page 1

H = 160837 = 165½ lbs on sq foot nearly = 10 tons 13 cwt on sq ft

Equivalent Pillar 213 feet high

No	Wt of each Stone	Wt of half Arch	Angle of the line of Pressure
1	2220	2220	0°·47
2	4749	6969	2°·29
3	4877	11846	4·13
4	5648	17494	6·13
5	6003	23497	8·18
6	6603	30100	10·36
7	6951	37051	12·58
8	5564	42615	14·50
9	5998	48613	16·49
10	7196	55809	19·8
11	7406	63215	21·27
12	10190	73405	24·32
13	28208	101613	32·17
14	22499	124112	37·39

Small Arch H = 20520

No	Wt of each Stone	Wt of half Arch	Angle of the line of Pressure
1	3000	3000	8°·19
2	8374	11374	29·0
3	8704	20078	44·23
4	16410	36488	60·39
5	20068	56556	70°·3

W of Large Arch = 124112
Small = 56556
Part of Pier = 12741
Sum = 193409

H in Large Arch = 160837
Small = 20520
Diff = 140317

Hence 54°2 = vertical angle of the resultant
Resultant = 83 tons 1 cwt for 1 ft in breadth

Fig. 25. Arch analysis for Maidenhead

A drawing of it is still available at Paddington. This had an intrados which was an arc of a circle of 110 feet radius but otherwise, in general concept, was superficially similar to Brunel's other arch bridges. Longitudinal voids occurred in the spandrels which had longitudinal ribs. The arch ring was 3 feet deep at the centre line. From these drawings the distribution of weight over the span can be estimated and from this the simply supported bending moment obtained. (This is sensibly what Brunel did by deriving and integrating his polynomials.) A value of the horizontal thrust H can then be chosen so that the relieving moment, namely H times the vertical height of the arch, reduces the net moment to any desired value. The author found that the arch shape is such that the thrust line cannot follow the arch centre line as closely as is the case with Brunel's calculations for the Maidenhead, Thorney Broad and Avon bridges.

With the Somerset bridge, Brunel again kept the centering in place as long as he dared and it may be that continued movement of the bridge relative to the centering and calculations like those just described, but of which there is no record in the manuscripts, resulted in the following being MacDermott's account of Brunel's replacement of this bridge:

> "This part of the railway had not been open long when a public outcry arose at the obstruction to the River caused by the centering of the Somerset Bridge, which, owing to a slight movement of the foundations, Brunel had not ventured to remove. At last the Directors were obliged to insist, and in August 1843 Brunel reported: With regard to Somerset Bridge, although the Arch itself is still perfect, the movement of the foundations has continued, although almost imperceptibly, except by measurements taken at long intervals of time; and the centres have, in consequence, been kept in place. Under existing circumstances, it is sufficient that I should state in compliance with a Resolution of the Directors, measures are being adopted to enable us to remove these centres immediately, at the sacrifice of the present Arch.
>
> "Six months later the Directors stated that 'a most substantial bridge has been built over the River Parrett without the slightest interruption to the traffic.' Between the same abutments Brunel had substituted the timber arch which did duty till 1904, when the existing steel girder bridge was erected."

The closeness of Brunel's thrust lines and the centre lines of the arches for the Maidenhead, Avon and Thorney Broad bridges suggests that a deliberate design approach to achieve this might have been developed, along the following lines (at least this might be expected from the advice given to one of his assistants in a letter dated 30th

December, 1854, in which Brunel writes, "... all forces should pass exactly through the centre of any surfaces of resistance ... especially ... in anything resembling a column or strut" (I. Brunel, 1870)).

In the initial stages of a design the line of rail and the position of the springings would be available. From these the approximate span and the rise of the arch would be found. Then from the many rules available or Brunel's experience the thickness of the arch ring at the crown would be chosen. With an allowance for filling over the crown, the intrados of a possible arch could now be sketched and starting from the extrados at the crown, the extrados of the arch also sketched. This would probably give an arch ring, increasing in thickness towards the roots. The positions where it became practicable to start the longitudinal walls to lighten the spandrels would now become apparent and a possible bridge would thus be available in outline. Division of this by vertical lines into, say, twelve sections would give the weight distribution over the span, whence from an appropriate polynomial the horizontal thrust and a possible thrust line for the arch could be obtained, starting from the crown and choosing the thrust line to be at mid arch ring depth there and also, say, at quarter span. The ring shape could now be adjusted in depth so the thrust line would be close to half ring depth over the whole span. In the Maidenhead bridge (Fig. 24) this might have been accomplished by just changing the starting point for an extra ring of bricks in the arch. The shape of the intrados would remain unaltered and there would be no difficulty (on paper) in altering the ring depth in this way. It would probably result in a small change in the weight distribution over the span and it may well be that Brunel's calculations in the manuscripts were his final check to ensure that the thrust and arch centre lines did not deviate appreciably.

The rapidity with which such a process of design converged would depend on the skill of the designer, particularly in choosing the initial shape for the arch. In the case of the Somerset bridge it may well be that this process was not used. If it had been, the form of the intrados near the roots would have been altered and the circular arch shape would have been modified.

This design approach is fundamentally sound in that it is an endeavour to produce an arch structure which is entirely in compression and so potentially a Maxwell (1869) optimum structure. The arch of minimum weight of material (cf. Owen, 1965) would have a considerably greater rise, but in the main Maidenhead arches the choice of rise seems to have been limited to the small value Brunel was forced to use by the terrain and line of rail. The design approach outlined is simple and avoids the complications and design tables which have not been a very sound or desirable feature of many of the texts on arches which have appeared from time to time.

W. H. Barlow's paper of 1846 and the discussion following it contain comments which may be of relevance in relation to the trouble which arose when the centering of the Maidenhead bridge was eased in 1838. "Lines of thrust", writes Barlow, "must not be brought too near the extremities of the voussoirs and in brick arches, particularly those turned in separate rings, a much greater latitude must be allowed." Barlow also emphasised that it was desirable "to form brick arches as much as possible in one bonded mass, using the best cement", and Bidder reinforced Barlow's remarks by commenting that Barlow's paper "showed the impropriety of constructing brick arches in separate super posed rings: the line (of thrust) would in almost every instance, travel out of the ring in which it commenced and in case of fracture the rings would fail consecutively but if . . . well bonded . . . would be more satisfactory. All the best arches were now built with full bond." Brunel, who was present at this meeting, made no reference to bonding, although his father had done work on this as indicated by Rennie (1846). Brunel chose to emphasise the need for arches to be considered as elastic homogeneous bodies (a study which was pursued later both in this country and abroad and found to be appropriate and relevant provided the nature of the support at abutments is known).

Fowler some years later widened Brunel's Maidenhead bridge by adding on each side arches which follow the profile of Brunel's bridge, but the arch ring depth is only that of the edge of Brunel's original arch. The aesthetic effect so obtained is to suggest that the arch ring is more slender than it actually is. Fowler used "superior bricks" (a needless compensation for the reduction in mean arch ring depth), "Portland cement and clean Thames sand possessing a degree of cohesive strength and regularity in quality . . . unknown in the days of lime and Roman cement" (Mackay, 1900). It is significant that Fowler did not change the shape of the intrados so he must have been satisfied with whatever he considered to be the position of the thrust line, but in spite of thinking that the piers were on "rock foundation", when he wrote his letter of 11th May, 1838, his foundations were extensively piled. He too would have known that it was essential to ensure that the abutments provide the necessary horizontal thrust. Isambard Brunel (1870) tells us that it was "originally intended that the foundation of the bridge should be on the chalk, which was at a short distance below the surface; but it was found to be very soft, and Mr Brunel therefore decided to place the foundations of the bridge on a hard gravel conglomerate overlying the chalk." The centre of the arches on the landward side of the main arches, he continues, were loaded with concrete and:

> ". . . were struck, and an active thrust opposed to the main arches before their centerings were eased. The line of pressure of each main arch was diverted downwards by the thrust of the flat arch adjoining it without the necessity of employing a great mass of brickwork in the abutment."

It thus appears that the following might well be an explanation of what happened to the Maidenhead bridge when the centering was eased. The western span stood firm because the mortar in the joints had set sufficiently firmly and the western abutment did not effectively move laterally. In the eastern span the "Roman cement" may not have had adequate time to set; the contractor, according to MacDermott (1927) and Rolt (1957), seems to have accepted this, although Brunel states (according to MacDermott) that the contractor considered that the abutments had moved. Some spanwise outward movement of the eastern abutment and the squeezing-out of mortar, which had not set between bricks, could both have had the effect of reducing the horizontal thrust until it reached such a value that the thrust line at the crown approached the outer ring in which the arch was turned. If this was not bonded adequately, it could have failed as a bent strut not bonded to the ring next to it, i.e. by the crown of the arch moving upwards. The next ring down could then have failed in the way described by Bidder (W. H. Barlow, 1846) as the mode of failure of an unbonded arch. (Was he describing what he thought to be the nature of the Maidenhead bridge failure?) Thus arch rings may have buckled upwards at the crown in turn and this process resulted in the arch settling on the centering "but not at the crown" just as Fowler describes (Mackay, 1900).

It could have resulted in the bottom three courses of bricks separating to "the extent of half an inch" for 12–15 feet on either side of the crown (Rolt, 1957). In the meantime rotation could have taken place about a fourth "hinge" near the intrados near the springing at the abutment where Fowler reported cracking "from top to bottom". The mortar in the longitudinal ribs would not have set to provide the added strength, bond and stiffness which Ruddock (1974) correctly indicates to be the advantage of this form of spandrel lightening.

From the satisfactory performance of the western arch Brunel knew that the basic design of the similar eastern half would be satisfactory if it could only develop the necessary horizontal thrust. Maintaining the centering in place was an insurance against disaster and allowing the bridge to continue in use to convey spoil to the western embankment would have retained some load on the arch ring and the abutments and helped consolidation, and on 8th October (MacDermott, 1927), when the centres were again eased, there was no further trouble. But in view of the earlier trouble a wise engineer would have retained the eased centering in place and used measurements taken between it and the arch to check whether further settlement was taking place. This would not be, as Rolt (1957) suggests, "excessive caution" or indicate "an impish sense of humour". The advice Brunel passed on to his assistant was "that you cannot keep centres or shores up too long" (I. Brunel, 1870).

In the case of the Somerset bridge (MacDermott, 1927), Brunel indicates that careful measurements of its movement were taken, possibly from the centerings which he again kept in place. This time slight movement persisted and, although the masonry arch appeared satisfactory, since he was being compelled to remove the centering he prudently replaced the bridge, presumably lest continued outward movement of the supports would eventually result in the horizontal thrust not being large enough to ensure the safety of the arch.

VI

Timber Works

DR L. G. BOOTH

It can be argued that Brunel was the greatest timber engineer this country has ever known. The coming of the railways presented challenges of a scale never before experienced by engineers, and of all the railways engineers he was the one who turned to timber most frequently and used it with the greatest flair and confidence. Brunel's exploits with timber should be seen against the rapidly changing background of knowledge and the materials that were available to structural engineers. After centuries of dependence on masonry and timber, the advent of cast iron as a structural material at the end of the eighteenth century apparently offered the new generation of engineers a material of immense potential. Telford was quick to seize the opportunity and, at the time, his lead must have suggested that timber would be rapidly relegated to become a secondary material only fit for house roofs and floors. And yet in 1844, some 60 years after the building of the iron bridge at Coalbrookdale, Brunel was working on a project for a timber railway bridge with a span in excess of 250 feet. Where other engineers were quick to use cast iron, Brunel mistrusted the material and it was probably this view, together with the high cost of wrought iron, that first led him to use timber and, with his increasing confidence, to accept it as a major structural material.

Before considering Brunel's work in detail, it is appropriate to mention the problems that faced (and still face) the engineer wishing to design large timber structures.

First, he needed reliable data on the strength properties of timber and, if the timber was to be used externally, on the methods of preventing decay. It appears that Brunel carried out detailed experiments on both topics, but little information on his work is now available.

Secondly, he needed to develop structural forms able to withstand greater loads, often over longer spans, than ever before. He faced the age-old problem that the structural use of timber is inevitably governed by the cross-section and length of the pieces that are available from the tree. Engineers have striven for hundreds of years to eliminate these limitations and two main methods have been developed. The first method is to form a framework in which the individual pieces of timber are arranged in a pattern that distributes the loads to the supports; an example is the truss with its top and bottom chords and intermediate bracing members. The second method is to take individual pieces of timber and to fasten them together continuously to form a component which is not limited by the size of the individual pieces; an example is a beam in which two pieces, known as laminations, are bolted together throughout their length to form a beam of double the depth of the individual laminations. Brunel made use of both methods and as this chapter develops it will be difficult to argue that he developed any completely new structural form.

Both methods require efficient jointing techniques. In the truss, joints are required at the node points and they must generally be capable of transmitting tension or compression forces. Efficient tension joints in timber were not developed until the twentieth century, and there is no indication that Brunel tried to solve this problem. He was content to use timber primarily in compression and to use wrought iron for tension members; in this way he generally avoided jointing problems. In the laminated component the joints between the faces of the laminations must resist the shear forces which cause the laminations to slip at their interface; for this essentially practical problem Brunel was able to develop a satisfactory solution.

The theoretical extent of Brunel's work is dealt with in Chapter IX and only a brief comment related to timber is needed here. His papers at the University of Bristol and in the British Transport Commission Archives give the impression of a practising consulting engineer rather than a research engineer. Although he performed numerous tests throughout his life, the emphasis was on testing for some particular end use rather than for the purpose of developing a general theory. According to William Bell, quoted in the biography by I. Brunel (1870), "he scarcely ever made any large girder or framework without having it fully tested, and he made extensive and elaborate experiments, most of them on a very large scale, on the strength of some of the materials and component parts of his different structures." Following Brunel's example, this chapter begins by considering his tests to find the physical properties of timber and then goes on to look at his timber roofs and bridges.

In one of his notebooks of facts (British Transport Commission Archives, *BTCA, 1*) which were compiled in the 1830s, he copied out the strength properties of various species as determined by Belidor (1739), P. Barlow (1817) and Tredgold (1820). He also recorded the formulae for strength and deflection and painstakingly copied the usual rules by which these formulae were transformed into words. Later he performed tests of his own to determine the deflection of large cross-sections in bending. Of a more theoretical nature were his tests to determine the position of the neutral axis in yellow pine. A recess was cut in the side of the beam at mid span and lined with hoop iron. A number of bars of very soft wrought iron with their ends reduced to points were placed horizontally and fixed hand tight in the recess. When the beams were loaded the position of the neutral axis in the depth of the beam was found when those bars on the tension side fell out *(BCTA, 2)*.

Bell (in I. Brunel, 1870) mentions elaborate tests made in Bristol in 1846 to determine the compression strength of yellow pine. The specimens, which were from 10 feet to 40 feet long and from 6 inches to 15 inches square, were tested in a specially designed framework nearly 50 feet high. The load was applied through a hydraulic press and the compression of the specimen was measured on all four faces and its lateral movement on two faces. From a knowledge of the modulus of elasticity (previously found from bending tests) the actual behaviour was compared with that predicted by Euler's theory. The information obtained on the compression strength of yellow pine was "new at the time, and almost essential to Mr Brunel in designing the many viaducts which he afterwards constructed". Brunel also made experiments in compression perpendicular to the grain to find the necessary bearing area under the washers of bolts and to determine the safe load on sills of viaducts. From the brief description given by Bell, these tests appear to be a milestone in the history of strength testing of timber in this country. Unfortunately, the test records have not so far been located.

The problems of preservation of timber are worthy of greater attention than is possible within this chapter. One can, however, say that the merits of the various processes engendered as much ferocity of feeling as was produced by the protagonists in the gauge war.

Between 1737 and 1849 patents were taken out for 47 different processes (Burt, 1853). The most frequently used were those invented by Kyan (chloride of mercury), Burnett (chloride of zinc), Margary (sulphate of copper) and Bethell (creosote). Brunel was a firm believer in the need for preservation and performed numerous experiments to try to determine the efficacy of the different processes and to discover if they had any adverse effect on the strength of timber. In his early days he regularly kyanised timber as a protection against decay and fire, but towards the end of his career he preferred Burnett's process when the timber was under cover, and creosote for use out of doors: when "expense was not a material object, both processes should be employed". Despite his experiments and his extensive experience he sounds a disillusioned man when he states:

> "It would, however, appear, that but little was positively known of the relative value of the various preparations, or of the mode of using any of them." (Brunel, 1853a)

In the circumstances one can see why it was essential that "he also minutely attended to the details by which timber structures may be protected from decaying influences" (Bell, in I. Brunel, 1870).

When Brunel began his career, timber was the traditional material for roof structures and, although cast and wrought iron eventually replaced it for large span structures, it remained the foremost roofing material throughout his life. For the small spans necessary for the majority of railway buildings he was able to use king and queen trusses; these designs were readily available to him in the standard books on carpentry by Nicholson (1793, 1797) and Tredgold (1820). The functional requirements of engine sheds called for larger spans, and for the main station roof a combination of functional requirements and prestige demanded even larger clear spans. Brunel responded to these challenges by using essentially functional roof trusses for the engine sheds and by adding decoration when they were used for the roofs of the train sheds.

The engine house at Swindon and the goods shed at Bristol are examples of Brunel's functional design at its best. The engine house was 290 feet long by 140 feet broad, with the breadth divided by two rows of columns into three compartments. The central part, which had a 50 foot span, used timber rafters and wrought iron tension members. The goods shed (Fig. 26(a)) had a central span of 60 feet with two side spans of 40 feet each. The columns were placed at 35 foot centres along the building and longitudinal trusses were provided in addition to the transverse trusses. Both sets of trusses used timber compression members with wrought iron ties. The combination of the solid-looking timber members and the light metal ties created roofs that were both functional and elegant. Both roofs were illustrated by Bourne (1846) in his description of the Great Western Railway.

At Bath and Bristol Brunel built major station roofs and in both cases he apparently felt it appropriate to obscure the structural action of the roofs by the addition of decorative timber work. Of the two, Temple Meads at Bristol is the more obscure and misleading. The roof is still in use although it now suffers the indignity of providing the cover for a car park. The roof is often described as a hammer-beam, and so it appears both today and in Bourne's drawing (Fig. 26(b)). In fact, the structural action is quite different and the hammer-beam effect is decorative.

In one of his sketch-books *(SB, GWR 6)* Brunel shows a similar-looking roof (without the pendants) and it may be that the thickening at the haunch was a misguided attempt to eliminate the thrust at the top of the walls. At what stage he realised that this solution was impracticable is not certain, but the drawing now held by British Railways at Paddington shows a tie bar incorporated at the level of the top of the pendants. Presumably Brunel thought that this design lacked the elegance required for his western terminus and it was abandoned. The final design created great interest as can be seen from Bourne's (1846) description:

> "The roof is particularly well suited to the purposes to which it is applied, and as it covers a clear span of 74 feet, without the aid of either cross tie or abutment, a particular notice of it will not be out of place here. It is composed of a series of 44 ribs, 22 on a side, and placed 10 feet apart, each of which is constructed somewhat like the jib of a crane, that is to say, of a long arm, projecting far and rising high into the air, and a short arm or tail, which in heavy cranes is either tied down or loaded. In the present case, the iron columns which divide the central space from the aisles are the fulcra or crane posts upon which the arms rest. The long arm or jib extends to the centre and ridge of the roof, and there meets its fellow from the opposite side, while the short arm or tail is carried backward to form the roof of the aisle, to the outer wall of which it is held down by a strong vertical tie passing some way down. The whole is then planked over diagonally, and is intended to be filled up, and decorated to suit the rest of the building."

Bourne's description can at least be accepted as structurally accurate. There is, however, a confusing post-script by Brunel himself. At a meeting of the Institution of Civil Engineers the Earl of Lovelace (1849) described the construction of a series of 24 foot span

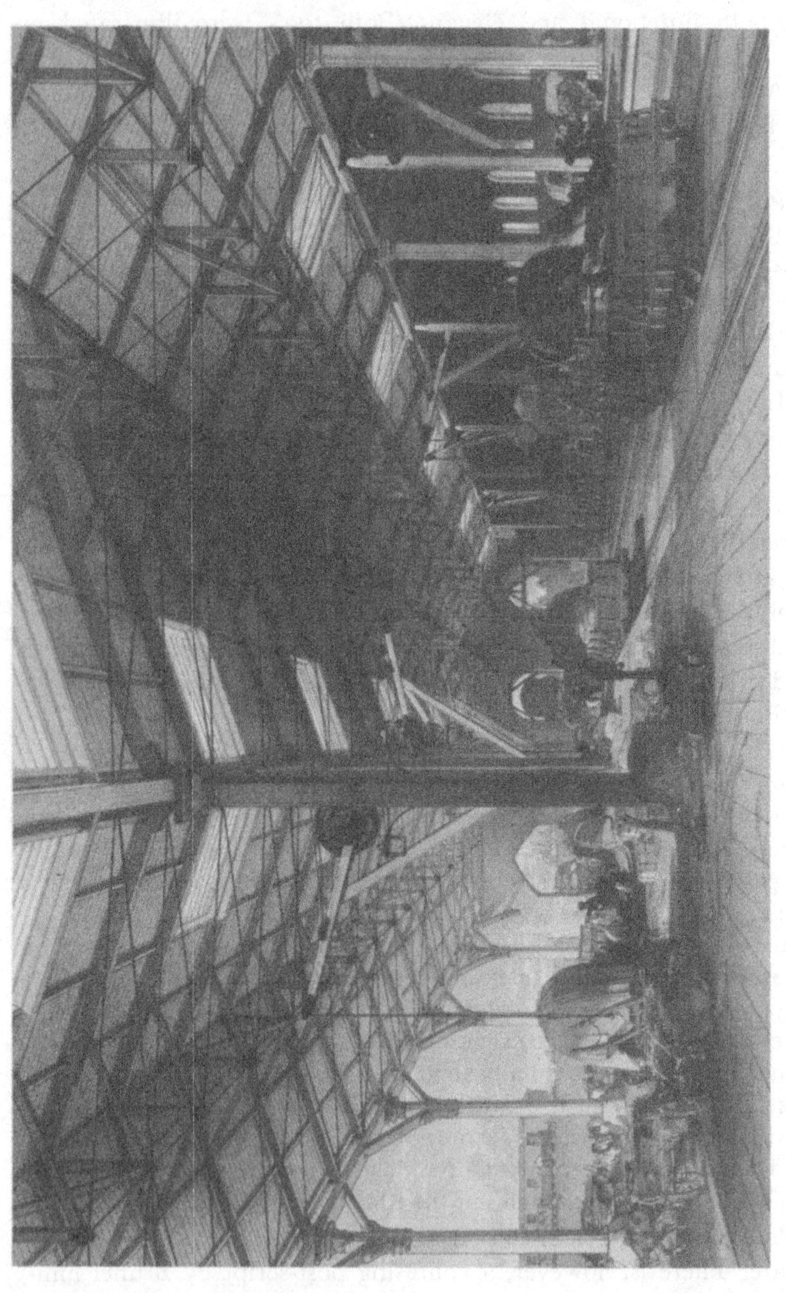

Fig. 26(a). Goods shed, Bristol

Fig. 26(b). Interior of Bristol station

laminated arches for the roof of his new hall at East Horsley. The arches were formed from 3 inch thick laminations steam bent to the shape of a pointed Gothic arch with vertical legs supported on corbels some 8 feet from the top of the walls. The laminations were moulded on their faces before they were steamed. In the discussion on the paper Brunel expressed admiration for the roof and then went on to mention Temple Meads. He considered his roof was an improvement on the ordinary system but he preferred the Earl's solution. Temple Meads was:

> "... formed by framing timber together, plating them on both sides with iron, and disguising the construction by ornaments applied to the timbers, instead of working them in the timber, as had been done by Lord Lovelace. In fact he had seldom seen so simple and useful a roof, possessing such an amount of stiffness, and at the same time avoiding all thrust upon the walls." (Brunel, 1849a)[34]

In some people's eyes the whole station at Temple Meads, and the roof in particular, represents a triumph of Victorian architecture and engineering. An alternative view would be that Brunel's architecture lacked originality and that it was in his engine and goods sheds and in his bridges that he produced his finest works in timber.

Although the main line of the Great Western Railway from London to Bristol ranks high among Brunel's engineering achievements, in terms of timber bridges the period was a formative one during which he steadily gained experience, and it was only on the completion of the line that he began to make much greater use of timber as a structural material. It is doubtful if this change can be attributed to one particular cause, but it may be due to the success of the station roofs at Bristol and Bath, the engine sheds and more particularly to the bridges at Sonning and Bath, which are examined later.

There are two possible ways of recording Brunel's bridges: chronologically or by structural form. Ideally the bridges should be examined chronologically, so that the development of his skills as a timber engineer can be traced, but for this method the actual dates of the designs rather than the dates of opening of the lines would be needed; a study of the site conditions would also be necessary to judge the possible structural solutions in the light of his previous and later designs. This would be an ideal solution, but unfortunately the necessary data are not always available. The recording by structural form, although less satisfactory, at least has the merits of an orderly presentation.

As we look at each type in turn it will be seen that a strong characteristic was the use of the material in large cross-sections with the result that the hazards of decay and fire were reduced: equally important, the strong, bold designs that were evolved avoided intricate details and minimised skilled workmanship. They were not large-scale joinery but were essentially carpentry, or what would now be called timber engineering.

The most frequent structural use of timber in the building industry is as a simple beam, but the need for larger clear spans often precludes its use in this simple form. The railways, however, brought a new problem: not only did they frequently require larger spans, but also the loads they imposed on bridges were greater than any previously encountered. Consequently, simple timber beams were rarely used as the primary structural element of a railway bridge. Their main use was as secondary members spanning the space between the primary structural members, which were often laminated beams and arches or various types of timber truss. A typical example was the use of closely spaced 9 inch thick timbers for the deck which carried the load from the rails to the longitudinal primary members. There is nothing unusual in Brunel's use of timber in this way; indeed it was a standard solution adopted by all railways not only in the days of timber bridges but often later when the primary structural members were iron or steel. The problem was never one of the adequate strength of the timber; the major difficulty was stopping decay in the damp conditions under the ballast, and to a lesser extent the ever present fear of fire caused by a flying spark.

In order to overcome the limitations on spans of bridges based on simple timber beams, Brunel began to experiment with laminated timber beams. If two pieces of timber of equal depth can be completely fastened together throughout their length to create a beam of double the depth of each piece, then it can be shown theoretically (and Brunel was aware of this) that the new beam will have a strength capable of spanning twice the distance of the single piece. The difficulty was to ensure that the laminations did not slide past each other as the beam deflected; today this would be achieved by gluing the laminations together, but the only adhesives available to Brunel were not waterproof and the bond would have quickly deteriorated in service. During August and September 1841 Brunel *(BTCA, 3)* investigated a number of different methods of fastening the laminations. In his first experiment he loaded two 20 foot span yellow pine beams of cross-section 14 inches square and recorded their deflections under increasing load. Two similar pieces were then fastened together with bolts passing vertically through the laminations and placed at 2 foot centres along the beam. Vertical marks were made on the side of the beam at the interface of the laminations and as the load was increased their relative movement "indicated that the slipping was gradual throughout the entire length of the beams". When the load was removed it was found that each beam had a permanent deflection and there was permanent slip between the laminations.

In an attempt to reduce the slip Brunel then inserted $1\frac{1}{2}$ inch diameter round dowels of American elm in the interface; two dowels were placed in each space between the bolts. Although this reduced the deflection of the beam, the dowels crushed into the laminations. In the next experiment the dowels were replaced by 6 inch by 2 inch pieces of American elm placed in notches at 45 degrees to the plane of the interface. Finally Brunel tested two beams of 39 foot span formed from two 10 inch square laminations. The beams were pre-cambered with $3\frac{7}{8}$ inch and $3\frac{1}{4}$ inch deflections and in this experiment the bolts passed through 6 inch by $1\frac{1}{4}$ inch cast iron plugs which were set in inclined notches at the interface.

Brunel's report contains a table of comparative deflections for the various interface devices and he concluded that the inclined cast iron plugs were the most successful in reducing deflection. An examination of the working drawings of his later bridges shows that he also used inclined bolts at the ends of the beams and that he often used hardwood keys, usually oak, in actual practice. Bell (in I. Brunel, 1870) refers to the hardwood and cast iron keys as joggles and states that Brunel perfected the method by using wrought iron wedges to ensure a tight fit of the joggles in the notches.

When the spans were too large for the use of laminated beams, Brunel naturally turned to trusses and he was able to call on his wide experience with the various forms of framing that he had used for the roofs of buildings on the London to Bristol line. His philosophy was clearly illustrated in his evidence to the inquiry into the application of iron to railway structures. When simple girders could not be used:

> "I avoid the use of cast iron whenever I can . . . In all cases, where it is beyond a certain span of from about 35 to 40 feet, I should prefer using timber or wrought iron, or the two combined." (Brunel, 1849a)

In 1842, at Stonehouse on the Bristol and Gloucester Railway, Brunel used 50 foot span queen trusses with timber bottom chords. With this arrangement it was possible to support the timber deck on the bottom chords and run the track between the trusses. For larger spans, such as the 74 foot St Mary's viaduct on the Cheltenham and Great Western Union Railway, the difficulty of making the heel joint probably forced Brunel to adopt a wrought iron bottom chord. Bell (in I. Brunel, 1870) describes the framing as a king truss with an internal queen truss (Fig. 27 (a)). It can be seen that Brunel's joggles were used where the inclined top chord was laminated from the heel joint to the horizontal top chord of the queen truss. The deck was carried on the top chord of the queen truss and on laminated beams which continued at this level to the supports; ten side spans of 20 feet also used laminated beams. A similar truss of 66 foot span was used in the Bourne viaduct. These trusses emphasise the value of Brunel's experiments on laminated beams and his expertness in combining timber and iron to the best advantage.

The trusses at St Mary's and Bourne stretched that form of framing to its limit, and when Brunel came to design what was his largest truss he developed a different framing. The viaduct at Landore on the South Wales Railway was 1,760 feet long with 37 spans which varied in length from about 40 feet to the maximum central span over the River Tawe of 110 feet. The description given by Fletcher (1855) to a meeting of the Institution of Civil Engineers represents the most detailed account available of any of Brunel's timber bridges. For the 110 foot span Brunel produced a complex frame (Fig. 27(b)) which is described by Fletcher as "of the bowstring description, the struts which form a portion of a polygon, being arranged in two rings, or series, one inside the other, and braced by wrought-iron tie bars". How the forces in the individual members were found is not explained, but Fletcher quotes the working stresses for the wrought iron and timber. An example of Brunel's attention to detail is his choice of cast iron rather than wrought iron washers. Cast iron washers could be made larger (six times the bolt diameter) and their use avoided the problem of the tendency of thin wrought iron washers to become hollow at the centre under pressure and to fail to bind the timber; they could also be cast with a bevel for use with skew bolts and so avoid cutting into the timber. A thoughtful structural development by Brunel was to increase the lateral strength of the bridge by using the decking as a horizontal girder. Instead of merely spiking the planks as separate strips he inserted 2 inch diameter wooden trenails in the joints between the planks.

In later years it became common practice for the early timber bridges to be replaced by iron and steel girders, and occasionally by

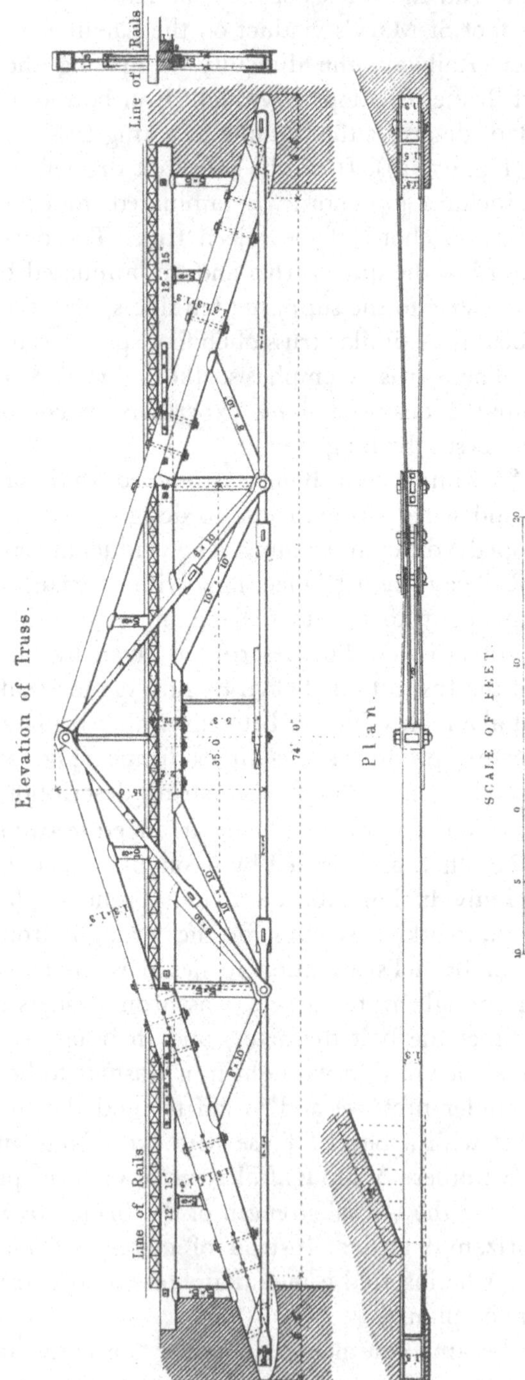

Fig. 27(a). Cheltenham and GW Union Railway: St Mary's viaduct

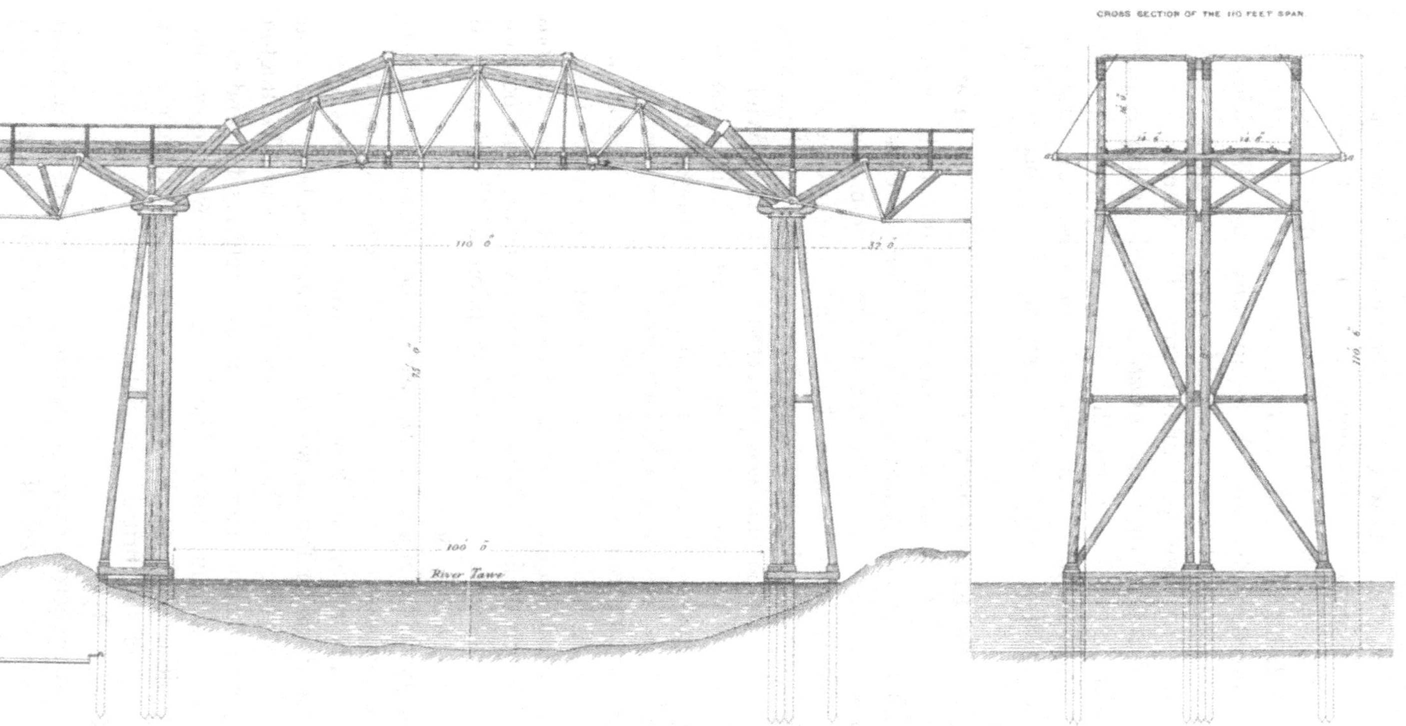

Fig. 27(b). Landore viaduct on the line of the South Wales Railway

masonry arches. Once, however, timber held the upper hand and replaced an unsatisfactory masonry arch. In 1841 Brunel built a masonry arch over the River Parrett on the Bristol and Exeter Railway. Although the arch itself behaved satisfactorily, by 1843 Brunel was forced to admit that the foundations were still moving and that a replacement would be required. He responded to this challenge by designing a 102 foot span timber arch. The framing was later described by Brereton (1871) as "a system of double polygons placed one within the other, bolted together and breaking joint, by which means great stiffness had been obtained." Only in 1904, after a life of 61 years, was it found necessary to replace the timber arch by a steel girder.

Although the River Parrett bridge was described as an arch it was only by the use of laminated timber that a true arch could be built. Laminated arches, in which the laminations were cut to profile and then notched or bolted to the adjacent lamination without being bent, were frequently used for bridges in Europe in the eighteenth century. The use of timber arched road bridges in which the laminations were bent into position was pioneered in Europe by Wiebeking in Bavaria during the years 1807–09. The same technique was used by Emy for various roof structures in France in the 1820s. There were many differences between Wiebeking's and Emy's structures but of major interest here is the fact that Wiebeking's laminations were usually about 12 inches thick and required considerable force to bend them to the required curvature, whereas Emy's laminations were only 3 inches thick and could consequently be bent into position without great difficulty. (A detailed description of Wiebeking's and Emy's work is given by Booth (1971a).)

In England the use of laminated timber arches for railway bridges was introduced by John Green. Green was Architect to the Newcastle and North Shields Railway and when the line was opened in June 1839 its major feature was the laminated timber viaducts at Ouseburn and Willington. Ouseburn had five arches and Willington seven; in both cases the spans were approximately 115 feet and each used 3 inch thick laminations. The bridges were well known and were the subject of a paper by John Green's son, Benjamin, at the 1838 meeting of the British Association for the Advancement of Science (Green, 1839).

There are numerous examples of Brunel's awareness of European practices and it seems likely that he knew of Wiebeking's and Emy's work; as a member of the British Association he was also probably aware of Green's bridges. Seen against this background of prior experience, Brunel's only laminated timber arch may seem to be of little importance, but his bridge at Bath did in fact have some small, but perhaps significant, differences.

In 1840 construction on the GWR from London to Bristol was far behind schedule, particularly so on the skew bridge over the Avon at Bath. Brunel had designed a two-span iron arch bridge, but there were difficulties over the tenders for the iron and he consequently decided to use laminated timber arches (Fig. 28(a)). The bridge had two 89 foot spans; each span had six parallel ribs placed at 5 foot centres and each rib was formed with five laminations. The laminations, which were 6 inches thick, were of Memel timber and were fastened together with bolts and iron straps. The ribs sprang from cast iron shoes in the masonry piers and the thrust was resisted by iron ties. The spandrels of the four outside arches were filled with an ornamental cast iron frame which also supported the timber parapet. The inner spandrels contained cross-ties and struts (Bell, in I. Brunel, 1870; Brereton, 1871).

There are several interesting aspects to this bridge. Although Brunel probably built more timber bridges than any other railway company Engineer, the Bath skew bridge appears to be his only laminated timber arch. It may be that he had difficulty in bending the 6 inch thick laminations, which were double the thickness of those used by Green and later by Vignoles, Locke, Robertson and Valentine on other lines. All the laminated timber arches had a limited life (usually about 25 years), their main weakness being that they were too flexible: the laminations tended to separate under load and the ingress of water inevitably led to decay. It may be that Brunel foresaw these problems and that his choice of 6 inch thick laminations, together with the iron ties to resist the thrust, represented his efforts to create a stiffer structure. In Brereton's view, speaking, perhaps with hindsight, in 1871, the success of the Bath skew bridge could be attributed to the use of thick laminations well preserved by kyanising (Brereton, 1871). When the bridge was replaced in 1878 its life of 38 years was exceeded only by Valentine's glued laminated bowstring arch over the River Wissey for the East Anglian Railway (Booth, 1971b). It seems strange that Brunel built no other laminated arches, particularly as the form would have been ideally suited to the valleys of Devon and Cornwall that he later mastered with timber viaducts.

Towards the end of the 1840s Brunel began to design the via-

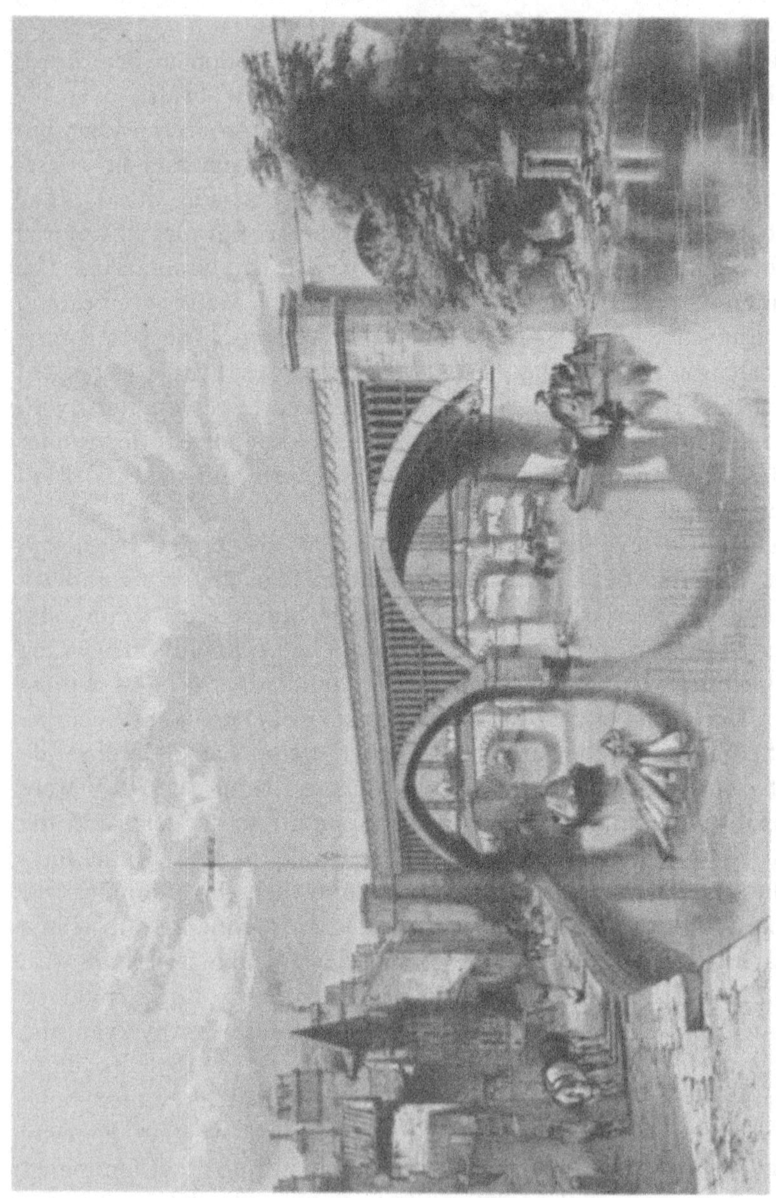

Bourne

Fig. 28(a). Oblique wooden bridge, Bath

British Rail

Fig. 28(b). Glazebrook viaduct

ducts that were to be his greatest achievement as a timber engineer. Trusses were occasionally incorporated in their construction but they are better described as horizontal continuous laminated timber beams supported on struts which fanned out from the top of masonry or timber piers. Although they will be mainly remembered for their use on the lines in Devon and Cornwall in the 1850s, their prototype can be traced to a bridge at Sonning which was built in 1840 to carry a by-road over the London to Bristol line.

When the Sonning cutting was complete Brunel was obliged to provide bridges to carry the public roads and for one of these he chose to use timber. The length of the bridge required was 240 feet and Brunel used timber beams supported on four timber piers. The piers, two at the bottom of the cutting and two on the slopes, were framed from large cross-sections of timber; they had a wide rectangular base and tapered to a line where they supported three longitudinal beams. Additional supports for the beams were provided by four sets of raking timber struts fanning out from the piers from about 12 feet below the road level. The road was carried on a timber deck which was supported on the three longitudinal beams. Since the loads were light there was no need for longitudinal ties between the piers.

Although Sonning contains the essential ideas of the later Cornish viaducts, Brunel's first solution to the problem of crossing the deep valleys of Devon was quite different. For five of the viaducts on the South Devon Railway he designed timber frameworks which were supported on masonry piers (Fig. 28(b)). Each frame, which was below the level of the railway, was part truss, part horizontal beams, and was described by Bell (in I. Brunel, 1870) as consisting "of a polygonal frame, with a few subsidiary struts, the feet of the main timbers being tied together by wrought-iron rods". The deck was carried by two frames, one at each side of the bridge. The bridge was originally designed when the Company intended to use the atmospheric system and was therefore not required to carry locomotives. When the atmospheric system was abandoned "a strongly trussed parapet was added above the trusses" to provide additional strength. The actual structural action is obscure and the result must now be considered as structurally unsatisfactory. The extent of Brunel's understanding of the theoretical behaviour of timber trusses is discussed in Chapter IX, but it may be significant to note that even before the strengthening he considered it expedient to test a complete span of the structure at Bristol before starting the erection. Each span was 60 feet; details of the length, height and date of replacement are given in Table 1. Several of the viaducts were built on a curve; outstanding were Ivybridge with a height of 114 feet and Blatchford with a length of 879 feet.

In terms of railway lines, the viaducts in Cornwall fall into three groups. The first line to open was the West Cornwall Railway from Truro to Penzance (25th August, 1852), followed by the Cornwall Railway main line from Plymouth to Truro (4th May, 1859) and finally the branch line from Truro to Falmouth (24th August, 1863). Although the last two lines were open at the end of Brunel's life and after his death respectively, the principles of the construction were evolved in the late 1840s and early 1850s (or indeed earlier at Sonning) and it is probably fair to consider them all as Brunel viaducts.

Although Brunel used timber for bridges more often than any other engineer in this country, his efforts in Cornwall make his previous structures appear infrequent. On the West Cornwall Railway he built ten timber viaducts on the 26 mile route; on the Cornwall Railway there were 34 timber viaducts in the 52 miles from Plymouth to Truro (the viaducts carried four miles of the line) and eight viaducts from Truro to Falmouth. The viaducts at Penponds, Hayle and Penzance had spans of only 20 feet and were formed with longitudinal beams supported on vertical timber piled legs; in a later reconstruction the timber piles were replaced by small masonry piers. These structures were not typical of the Cornish viaducts and were more akin to some of Brunel's small spans between London and Bristol. If these viaducts are eliminated, the Cornish viaducts may be classified as having two, three or four sets of raking legs springing from either timber or masonry supports; the legs supported longitudinal laminated beams which in turn carried the transverse decking. The dimensions of the viaducts varied considerably (Table 1) not only between companies but also along the individual lines.

The simplest form of structure, such as that at Penadlake (Fig. 29(a)), had 40 foot spans. Two raking sets of legs, with three legs in each set, sprang from small masonry piers which rose about 2 feet above ground level. The legs then supported three longitudinal laminated timber beams formed from two 24 inch by 10 inch pieces of timber. This form of construction could only be used when the valley was shallow and it was often chosen as being more economical than an embankment.

When three sets of raking legs were used (e.g. Angarrack, Fig. 29(b)) the span could be increased to 50 feet. The raking legs, which

Table 1. Brunel's viaducts in Devon and Cornwall

Location	Length in feet	Height in feet	Date of opening	Date of replacement
South Devon Railway				
Exeter to Plymouth				
Glazebrook	489	80	1849	1861–63
Bittaford	351	61	1849	1861–63
Ivybridge	756	114	1849	1861–63
Blatchford	879	107	1849	1861–63
Slade	819	100	1849	1861–63
Plymouth to Tavistock				
Cann	381	63	1859	1907
Riverford	381	97	1859	1893
Bickleigh	501	123	1859	1893
Ham Green	570	91	1859	1899
Magpie	648	62	1859	1902
Walkham	1101	132	1859	1910
Newton Abbot to Kingswear				
Longwood	600	40	1864	1921
Noss	510	40	1864	1921
Cornwall Railway				
Plymouth to Truro				
Stonehouse Pool	321	57	1859	1908
Keyham	432	90	1859	1900
Weston Mill	1200	46	1859	1903
Combe-by-Saltash	603	86	1859	1894
Forder	606	67	1859	1908
Wivelscombe	198	25	1859	1908
Grove	114	29	1859	1908
Nottar	921	67	1859	1908
St Germans	945	106	1859	1908
Tresulgan	525	93	1859	1897
Coldrennick	795	138	1859	1897
Treviddo	486	101	1859	1897
Cartuther	411	89	1859	1882
Bolitho	546	113	1859	1882
Liskeard	720	150	1859	1894
Moorswater	954	147	1859	1881
Westwood	372	88	1859	1879
St Pinnock	633	151	1859	1882

continued

Table 1—continued

Location	Length in feet	Height in feet	Date of opening	Date of replacement
Largin	567	130	1859	1886
West Largin	315	75	1859	1875
Draw Wood	669	42	1859	1875
Derrycombe	369	77	1859	1881
Clinnick	330	74	1859	1879
Penadlake	426	42	1859	1877
Milltown	501	75	1859	1896
St Austell	720	115	1859	1898
Gover	690	95	1859	1898
Combe St Stephens	738	70	1859	1886
Fal	570	90	1859	1884
Probus	435	43	1859	1871
Tregarne	606	83	1859	1902
Tregeagle	315	69	1859	1902
Truro	1329	92	1859	1904
Carvedras	969	86	1859	1903
Truro to Falmouth				
Penwithers	813	90	1863	1926
Ringwell	366	70	1863	1933
Carnon	756	96	1863	1933
Perran	339	56	1863	1927
Ponsanooth	645	139	1863	1930
Pascoe	390	70	1863	1923
Penryn	342	83	1863	1923
Collegewood	954	100	1863	1934
West Cornwall Railway				
Truro to Penzance				
Penwithers	372	54	1852	1887
Chacewater	297	52	1852	1888
Blackwater	396	68	1852	1888
Redruth	489	61	1852	1888
Penponds	873	45	1852	1899
Angarrack	798	100	1852	1885
Guildford	384	56	1852	1886
Hayle	831	34	1852	1886
Penzance	1041	12	1852	1921

now sprang from a level 41 feet below the deck, were supported on vertical timber legs which were in turn supported by masonry foundations. A continuous longitudinal timber member was placed at the level of the springing point of the raking legs; longitudinal strength was provided by this member and by iron diagonal braces between some legs. In 1869 masonry piers replaced the vertical timber legs and the size of the raking legs was increased.

The culmination of the structural form was the use of four sets of raking legs mounted on masonry piers at 60–66 foot centres (e.g. St Pinnock, Fig. 29(c)). The piers, which needed good foundation conditions, were about 7 feet thick with a batter of about 1 in 100; some of the piers were strengthened by being built in the form of a cross which thickened in steps according to the height. The tops of the piers were usually about 35 feet below the deck and it was from this level that the four sets of raking legs sprang from cast iron shoes. Each set of legs was transversely braced with horizontal and diagonal members. The three longitudinal beams were made from two 12 inch by 12 inch timbers, laminated with the usual raking bolts, joggles and keys. The deck was of 12 inch by 6 inch timber. Each bay was stiffened with metal ties which ran longitudinally and diagonally from the heads of the piers.

Where the line crossed tidal creeks, timber piles were used and the spans were usually reduced to 40 feet. At St Germans (Fig. 29(d)), however, the spans were increased to 66 feet and the raking leg construction was replaced by a timber truss with wrought iron tension members; this form was unique in Devon and Cornwall (Fig. 30).

The lengths and heights of the Cornish viaducts and their eventual dates of replacement are given in Table 1. Particularly impressive were the 1,329 foot long Truro viaduct and that at St Pinnock with a height of 151 feet. Fig. 31 illustrates two of these Cornish viaducts in their natural setting.

In some respects the choice of timber as the structural material for the viaducts had been determined by the lack of capital available to the companies. It was appreciated that timber viaducts would have higher maintenance costs than those built with more durable materials, but at least the initial cost was reduced and the line could begin to earn revenue. Some other companies adopted timber as a temporary measure and budgeted to replace it by masonry or iron girders when sufficient finance was available. There is, however, nothing to suggest that Brunel considered the timber viaducts to be

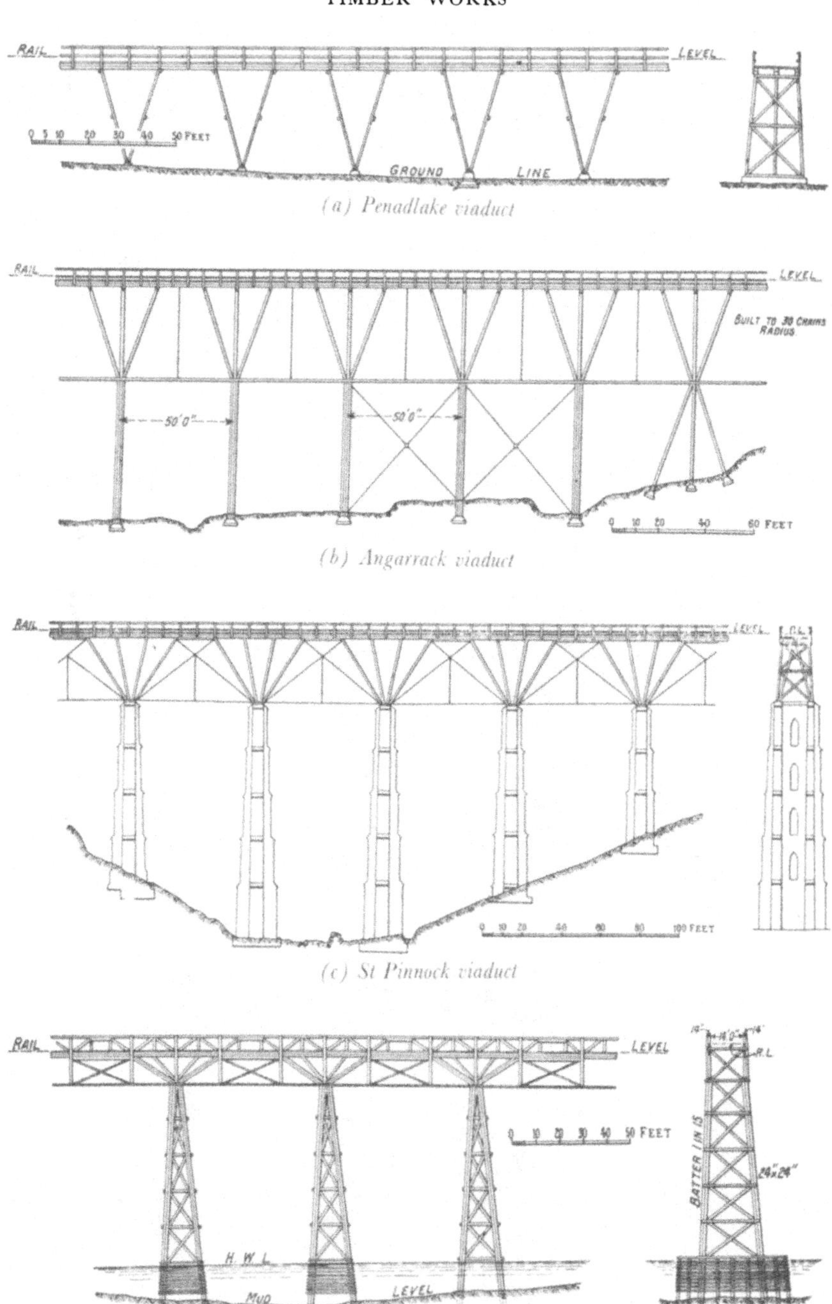

(a) Penadlake viaduct

(b) Angarrack viaduct

(c) St Pinnock viaduct

(d) St Germans viaduct

Fig. 29. Constructional schemes for timber viaducts

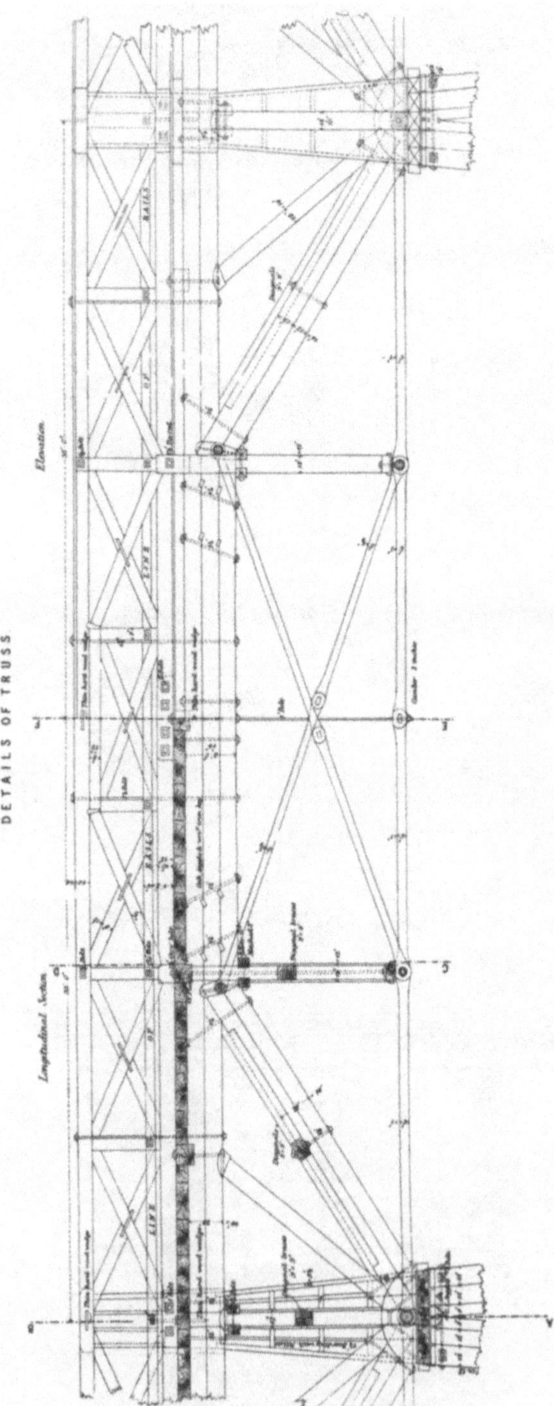

Tredgold

Fig. 30(a). St Germans Viaduct: details

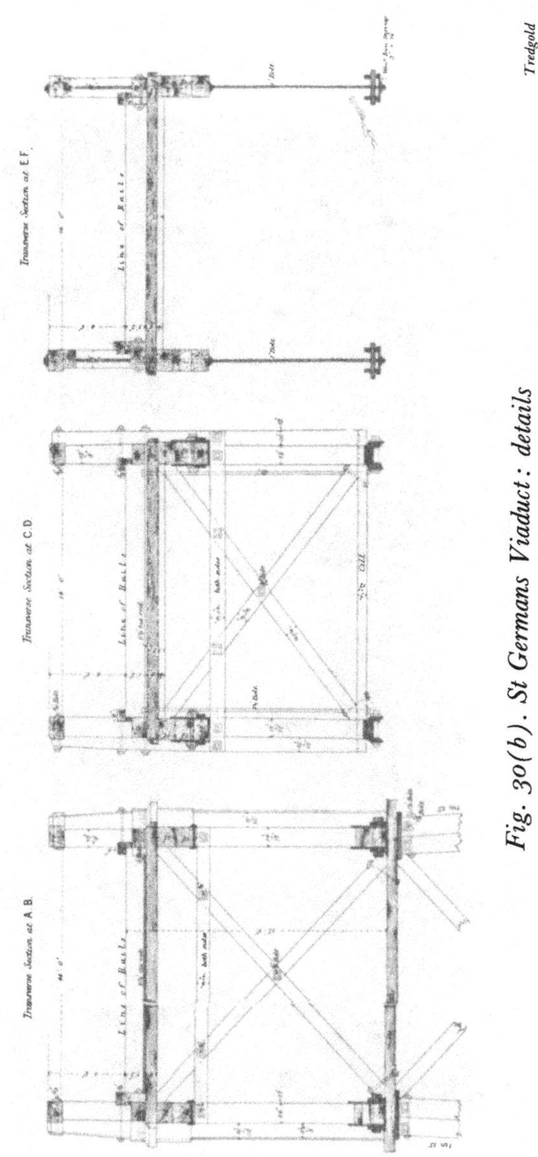

Fig. 30(b). St Germans Viaduct: details

Tredgold

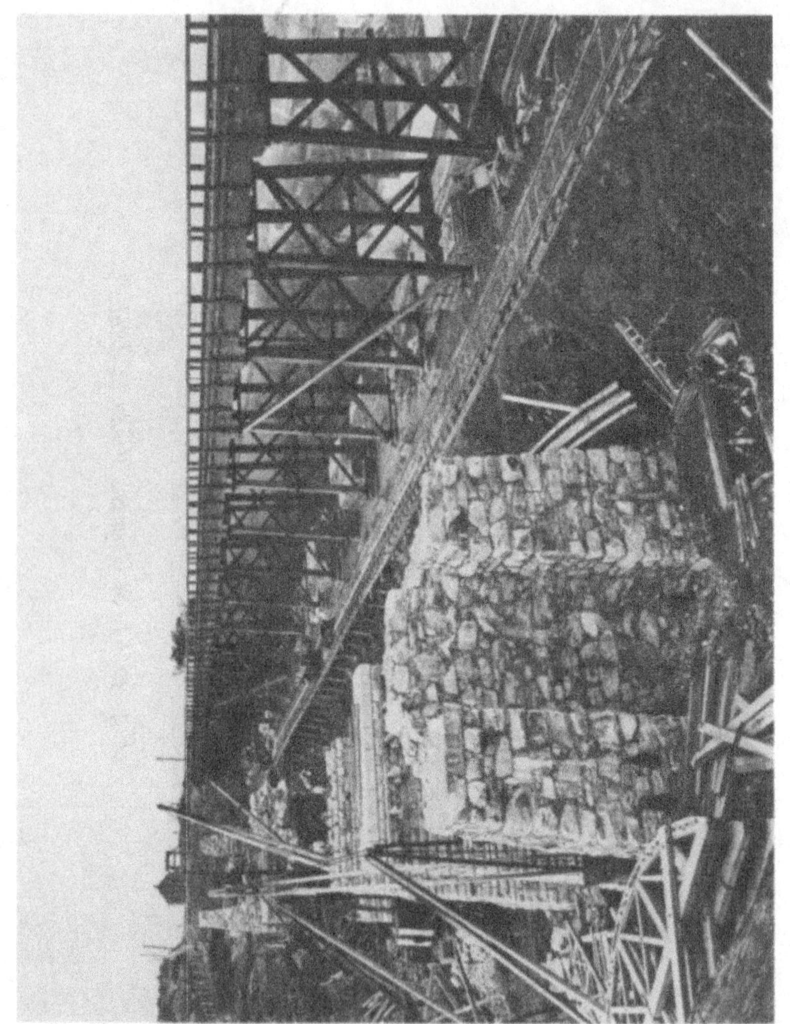

British Rail

Fig. 31(a). Penponds Viaduct

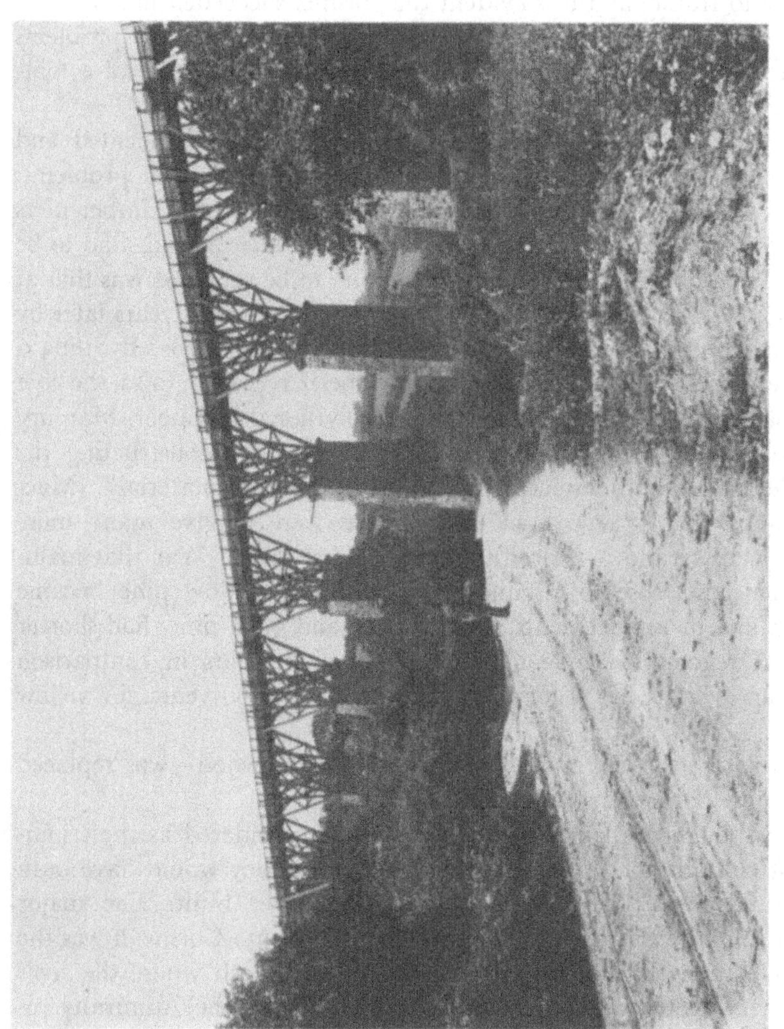

British Rail

Fig. 31(b). Gover Viaduct

of a temporary nature. Although the timber was kyanised Brunel considered that the yellow pine would have a limited life and he therefore paid special attention in the design to the problem of replacement of decayed members. Whitley (1931) describes in detail how the replacements could be rapidly inserted without having to close the bridge to traffic, and it is evident that Brunel succeeded in his objective; only at St Germans did the timber trusses cause any problems during maintenance and these were overcome by the use of a temporary truss.

Although the decayed members could be readily located and replaced, there were occasionally more serious structural problems. Some of the viaducts had been built on curves and the timber piers of those with insufficient strength to resist the lateral loads had to be reinforced with shoring. The first structure to be replaced was that at Probus (after a life of only twelve years), followed four years later by Draw Wood; both were of the Penadlake two-leg type. By 1883 a further nine viaducts in Cornwall had been replaced (also the five on the South Devon main line) and the Divisional Engineer, Margary "would strongly recommend the advisability of reconstructing the whole of the timber viaducts in a more durable material" (MacDermott, 1927). Several of the viaducts were to give many more years' service and it was only after the First World War that maintenance costs rose to an uneconomical level. Yellow pine became scarce and its replacements, Oregon pine and pitch pine, had shorter lives: Oregon pine gave a life of about eight years in comparison with an average life of 30 years, and even up to 60 years, for yellow pine.

In 1934 the last of the viaducts—at Collegewood—was replaced after a life of 71 years. An era had ended.

Although the Cornish viaducts can be considered as the triumphant culmination of Brunel's work in timber they would have been surpassed by a bridge which, sadly, he never built. The major obstacle of the railway's thrust through Devon into Cornwall was the River Tamar. Brunel planned to cross it at Saltash where the river was 1,100 feet wide. The restrictions imposed by the Admiralty required a bridge with a large central span and adequate head room. Brunel responded to the challenge with a proposal to build a bridge with one span of 255 feet and six of 105 feet "with superstructures of timber trussed arches" (Bell, in I. Brunel, 1870).

There seems to be no doubt that this was a serious proposal by Brunel and confirmation of his confidence in such a bridge can be found in a letter he wrote to Vignoles in November 1844. Vignoles had written to Brunel asking for his views on a large timber bridge he proposed to build on the line from Limerick to Waterford. In his reply Brunel was not prepared to pass an opinion on Vignoles' design, but he did say, "I have used timber a great deal in construction and see no difficulty in making arches for railways of 250 feet span—in fact I am projecting a larger one at the present" *(PLB, 22 Nov., 1844)*. This letter may have been referring not to Saltash but to Chepstow where he also proposed to use timber for the 300 foot span (Bell, in I. Brunel, 1870). In either case, the confidence was there.

When the Admiralty changed its requirements, the increased spans needed were beyond even Brunel's belief in timber and his proposals were abandoned in favour of the tubular trusses which were to be one of his greatest monuments. Sadly, no details of the timber design have yet been found.

The inadequacies of the preservation techniques then available would inevitably have led to decay and to the eventual replacement of the timber by a more durable material. But during its lifetime, what a triumph for Brunel, the timber engineer, it would have been.

VII

The Three Great Ships

PROFESSOR J. B. CALDWELL

In 1835, the year in which Brunel conceived the first of his three great ships, there appeared in a book on English trades the following passage:

> "The man of science and the practical shipwright have long lamented that, in the theory of the art of shipbuilding, there are so few fixed and positive principles established by demonstration, or confirmed by practice; thus the artist being left to the exercise of his own opinion, in general resists theoretical propositions, however speciously formed, so hard has it ever been found to overcome habitual prejudices. The great neglect of the theory of shipbuilding, is much to be deplored in a country like this, where the practical part is so well understood and executed." (Society for Promoting Christian Knowledge, 1835)

In naval architecture, where the evolution of arts and practices had so long preceded theoretical developments, "habitual prejudices" were inevitably strong. But the seventeenth and eighteenth centuries brought the beginnings of the science of naval architecture; and some understanding of the mechanical principles governing the behaviour of floating bodies was, by the early nineteenth century, well established. Thus Bouguer in France, Santacilla in Spain, Leonhard Euler in Russia and Chapman in Sweden had, among others, laid the foundations of the theory of ship stability and strength and had pioneered the experimental study of ship resistance. In Britain neither the art nor science of shipbuilding had advanced greatly in this period, despite—or perhaps because of—a flourishing demand for ships. Only Atwood, on stability, and Beaufoy, with his systematic experiments on ship resistance, stand out for their scientific contributions. There is little evidence, however, that their "theoretical propositions", any more than those of their continental contemporaries, were influencing ship designers in any significant way. Such treatises on shipbuilding as those by Murray (1765) and by Stalkartt (1781) (both of Deptford) and Gordon (1784) were devoted mainly to expositions of the shipwright's art, to the delineation and geometry of the hull form of sailing ships, and to the design of their rigs.

From the industrial ferment at the turn of the eighteenth century, technical innovations having possible applications to ships inevitably emerged. At last, practicable alternatives to wind for the motive power and to wood for hull construction could be recognised. Among the earliest British engineers to see the real possibilities of marine steam propulsion were Trevithick, Bell and Brunel's father, Marc, whose experiments with a small paddle-driven steamer on the Thames in 1814 did not, however, convince him of its practical application in ocean-going ships. Henry Bell's *Comet* (40 feet long, 10.5 foot beam, 4 foot draught, 4 horsepower) is generally credited with the inauguration of commercially successful steam propulsion in Britain, as a ferry on the Clyde, although American naval architects such as Fulton and Fitch had a few years earlier started similar operations on the Hudson river. During the 25 years following the *Comet*'s entry to service in 1812, the numbers, sizes and powers of steamboats for river and in-shore service rapidly increased. In 1837 the paddle steamer *Rainbow* (185 feet long, 25 foot beam, 600 tons, 180 horsepower) commenced operations between London and outports such as Bristol. Even the Admiralty, later to be so roundly castigated by Brunel for its "withering influence", ordered its own steam-driven *Comet* (115 feet long, 21 foot beam, 9 foot draught, 2 by 40 horsepower) for the Royal Navy.

The iron-hulled *Rainbow* was, in another sense also, a landmark in naval architecture and in its influence on Brunel. The substitution of iron for wood was proceeding almost in step with that of steam for sail. From John Wilkinson's canal barges (*circa* 1785), Jevons' Mersey boat (1815) and the *Aaron Manby* (107 feet long, 17.2 foot beam, 3.5 foot draught, having a steam engine fitted in an iron hull, 1821) to the establishment of Fairbairn's shipyard for iron vessel construction on the Isle of Dogs, the accumulation of experience ensured that progress was rapid, despite unwillingness in some quarters to believe that an iron hull could be made buoyant. More serious were the objections regarding its effect on the navigational compass, and questions about its durability and proneness to fouling. But to men such as the Lairds in Birkenhead, where they started iron boat construction in the early 1820s, and Fairbairn, a pioneer of iron for both ships and bridges, the material would undoubtedly eventually supersede wood, especially as suitable timber was by then becoming scarcer and more expensive.

The Industrial Revolution, clearly, was bringing major changes in marine industry. No less significant was the consequent growth of trading and transport which it helped to stimulate. This, then, was the general scene into which Brunel thrust himself by his proposal, in the autumn of 1835, to extend the new Great Western Railway to New York by building a steamship for service between Bristol and the New World. Of his subsequent contributions to naval architecture over the next 24 years, the bare technical facts are well known. Some principal particulars of his three great ships are given in Table 2.

Table 2. Some particulars of the three ships as originally built

Detail	*Great Western*	*Great Britain*	*Great Eastern*
Date of launch	1837	1843	1858
Length overall	236 feet	322 feet	692 feet
Length between perpendiculars	212 feet	285 feet	680 feet
Breadth over paddles	59.8 feet		118 feet
Breadth of hull	35.3 feet	50.5 feet	82.5 feet
Depth to upper deck	23.2 feet	32 feet	58.0 feet
Draught laden	16.7 feet	18 feet	30 feet
Displacement at load draught	2300 tons	3675 tons	27380 tons
Hull material	Wood	Iron	Iron
Iron weight in hull		1040 tons	6250 tons
Propulsion	Sails and paddles	Sails and screw	Sails, screw and paddles
Design horsepower	420	1000	Paddles 1000 Screw 1600
Paddle diameter	28.7 feet		56 feet
Screw: diameter		15 feet	24 feet
pitch		25 feet	44 feet
blades		6	4
revolutions per minute		54	39

Fig. 32 shows the three ships in profile, based on Baker (1965), Corlett (1971) and Russell (1865).

First came the *Great Western* (1836–56) of 2,300 tons displacement, developing 750 indicated horsepower for her paddle-wheels, substantially the largest steam vessel up to then and the first to demonstrate conclusively the economic and technical feasibility of transatlantic liner operations. Second was the *Great Britain*, floated out of a Bristol dock in 1843 and now happily restored in that dock after extraordinary and varied service, including the first screw-propelled crossing of the Atlantic in 1845 and 80 years as a storage hulk in the South Atlantic. In conception, technical innovation and performance, *Great Britain* has aptly been described (Corlett, 1971) as "the fore-runner of all ships of significance afloat today". Last came the *Great Eastern* (1858–88), having nearly three times the length and twelve times the displacement of the *Great Western*, propelled by sails, paddles and a screw, and subsequently considered variously to be "the most wonderful piece of naval architecture ever projected" (Elgar, 1894) or "a melancholy illustration of great ambition but great ignorance" (Harland, 1894).

We have seen that many of those "innovations" with which Brunel's ships are sometimes credited—for example, steam power, the use of iron, screw propulsion and ship subdivision (mentioned later)—were already being taken up, albeit on a smaller scale, when Brunel projected his first venture into shipbuilding. In point of size, however, he was a true pioneer, and it was in his enunciation and application of the broad principles of the "economies of scale" in ship size that he made his first and possibly most important contribution to the developing science of naval architecture.

It is not clear why, despite some clarification by Chapman (1775), Beaufoy (1834) and others of the nature of ship resistance, the opinion persisted that the coal required per mile to propel a steamship at a given speed would increase in simple proportion to her size, as measured by her displacement. On this argument, since no small vessel could be designed to carry enough coal to fuel a transatlantic voyage, neither could a large vessel accomplish this. Brunel's later notes show some familiarity with Beaufoy's work and with the notion that the resistance of a ship at a given speed depended primarily on the area of her largest cross-section. Erroneous though this view was in detail, it was closer to the truth, and convinced Brunel that the workings of the square–cube law pointed clearly to increased size as the only means whereby an economic ratio of fuel requirement

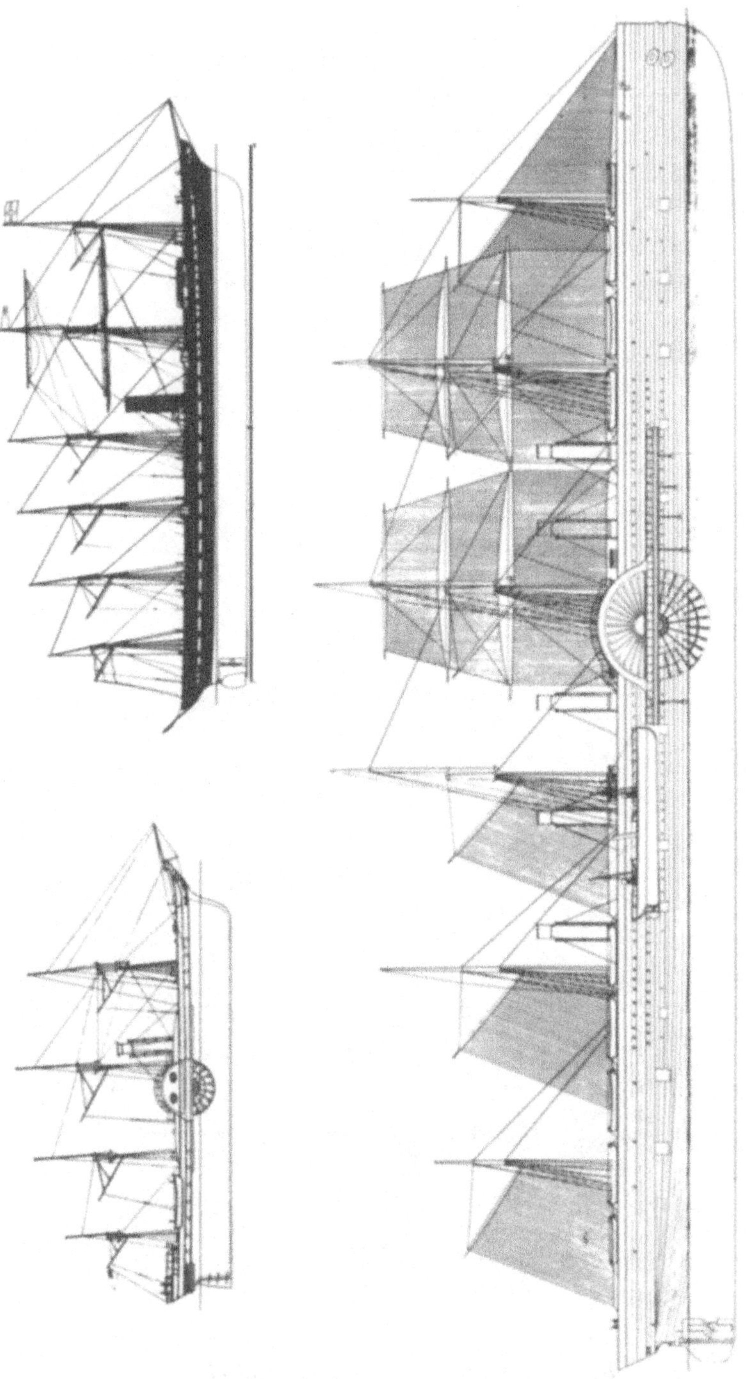

Fig. 32. The three ships in profile

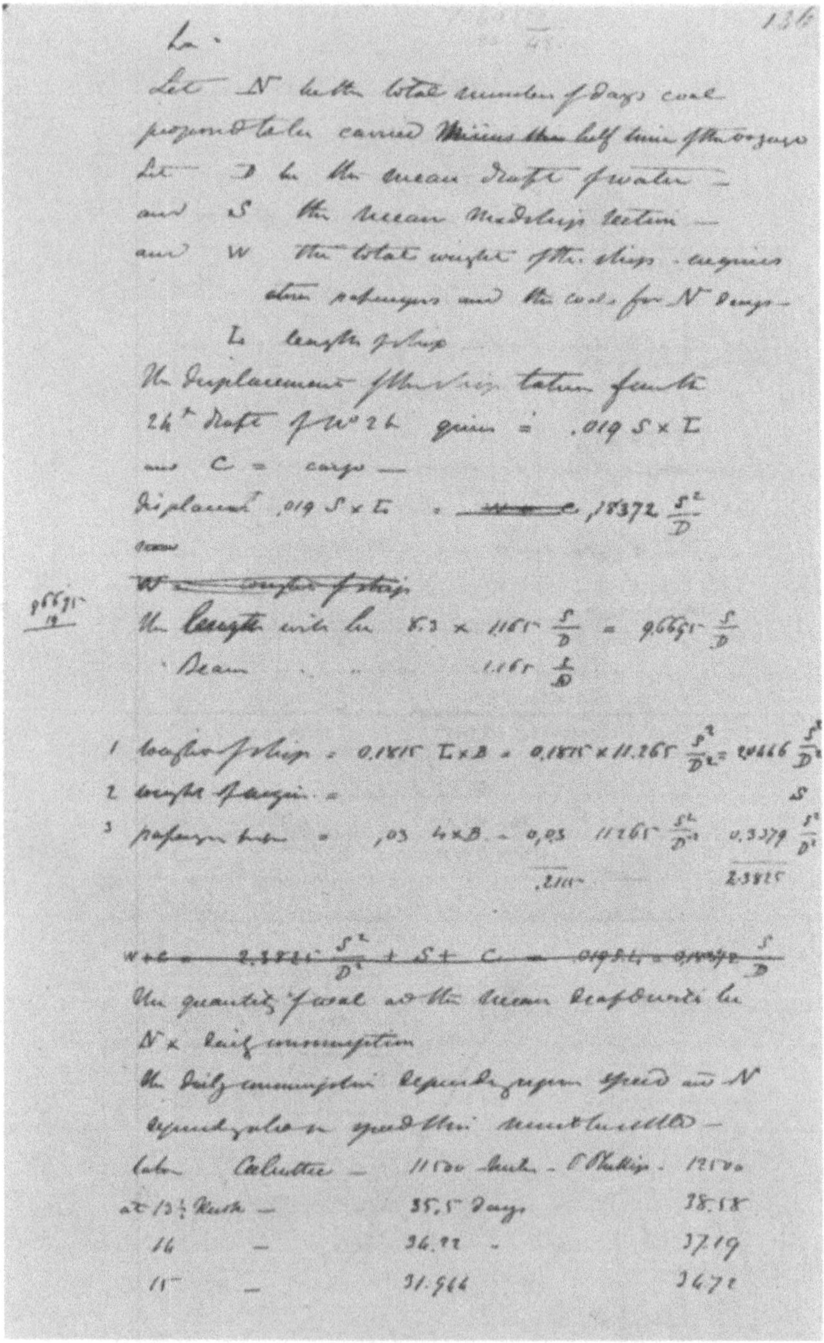

136

Let N be the total number of days coal proposed to be carried ~~minus the half time of the voyage~~

Let D be the mean draft of water —

and S the mean midships section —

and W the total weight of the ships . engines stores passengers and the coals for N days —

L length of ship

the displacement of the ship taken from the 24ft draft of No 26 given = .019 S × L

and C = cargo —

displacement .019 S × L = ~~[illegible]~~ .18372 $\frac{S^2}{D}$

~~W = weight of ship~~

9.6695 / 12 the length will be 8.3 × 1.165 $\frac{S}{D}$ = 9.6695 $\frac{S}{D}$

Beam . . 1.165 $\frac{S}{D}$

1 weight of ship = 0.1815 L×B = 0.1815 × 11.265 $\frac{S^2}{D^2}$ = 2.0446 $\frac{S^2}{D^2}$

2 weight of engine = S

3 passenger berths = .03 L×B = 0.03 11.265 $\frac{S^2}{D^2}$ 0.3379 $\frac{S^2}{D^2}$

.2115 2.3825

~~W + C = 2.3825 $\frac{S^2}{D^2}$ + S + C = .01951 [illegible] $\frac{S}{D}$~~

the quantity of coal at the mean draft would be N × daily consumption

the daily consumption depends upon speed and N depends also on speed then must be settled —

take Calcutta —	11500 miles —	C Phillips — 12500
at 13½ knots —	35.5 days	38.58
14 —	34.22	37.19
15 —	31.944	34.72

Fig. 33. Analysis of ship size: typical page

(dependent on ship sectional area) to fuel capacity (dependent on volume) could be achieved. In this reasoning Brunel had the support of John Laird who was no less contemptuous than he of the opposing views of authorities such as Dr Dionysius Lardner (1838), who wrote at length on the subject.

The method of analysis which Brunel used for his preliminary estimates of the dimensions of ships is well illustrated by the following edited extract from one of his calculation books *(CB, 1850)*. Fig. 33 shows a typical page.

"*FORMULA FOR SIZE OF SHIPS*

"First as regards consumption of coal . . . taking Slater's[35] formula

$$\text{if } V^3 = \frac{P \times 1200}{S}$$

V being miles (per hour), P being the indicator power of engine, and S the midship section,

and assuming $2\frac{1}{2}$ lb per hour per I.H.P., we have daily consumption

$$= \frac{2\frac{1}{2} \times 24}{2240} \times P = 0.0267 \frac{V^3 S}{1200}$$

$$\text{or if 3 lb per hour} = 0.0310 \frac{V^3 S}{1200}$$ [36]

"If V be taken in knots . . .

$$\text{daily consumption} = 0.0267 \frac{V^3 S}{700} \qquad \text{(No. 1)}$$

$$\text{or}^{[36]} = 0.0350 \frac{V^3 S}{700} \qquad \text{(No. 2)}$$

Taking V = $13\frac{1}{2}$	14	15 knots
No. 1 = 0.093831 S	0.104635 S	0.13070 S
No. 2 = 0.12300 S	0.137200 S	0.16875 S
Mean, say, 0.11 S	0.1200 S	0.15 S

"If the daily consumption be called c, the total cargo of coal . . . will be Nc, N being the number of days coal assumed to be carried.

"The displacement in tons of the ship will of course depend upon the beam, the length and the depth. The length will depend entirely upon the beam, and the beam will depend upon the draft assumed with a given capacity or displacement. For the forms of ships likely to be used in the present case for which only I propose to fit the formula, I shall assume the length to be 8.3 of the beam, and that the beam is equal to 1.165 of the midship sect. (sectional area) divided by the draft. This I find to be a mean result from the design no.2b taking the drafts of 20, 24, 28 and 30. Or the mid. sect. = 0.8584 of the beam by the draft.

"The weight of the hull will, I think, be in the ratio of the beam by the length—at least this seems to be the nearest and most convenient measure—because the total depth of hold will be nearly the same within moderate variations of the draft.

"If we want to be more precise and provide for a greater range we must take the draft also + a constant for the upper works, but the former will I think do.

"Taking the weight of the *Adelaide* as a guide at 900 tons, and assuming the depth of the ships now under consideration to be double that of the *Adelaide*, the weight of the ship would be 0.1815 × length × beam.

"I assume the weight of engines and boilers to be equal in tons to the M.S. (midship section) in feet—for such speeds as we here talk of.

"The passengers, ship's stores, crew and provisions I assume to be proportional to the deck area—assuming them within the range of the present enquiry there will be the same number of decks—and equal to 0.030 × length × beam.

"Let N be the total number of days coal proposed to be carried minus the halftime of the voyage

Let D be the mean draft of water
- S the mean midship section
- W the total weight of the ship, engines, stores passengers and the coals for N days
- L length of ship
- C cargo

"The displacement of the ship taken from the 24ft draft of No 2b gives $= 0.019\ S \times L$

$$\text{Displacement } 0.019\ S \times L = 0.18372 \frac{S^2}{D}$$

$$\text{The length will be } 8.3 \times 1.165 \frac{S}{D} = 9.6695 \frac{S}{D}$$

$$\text{Beam will be } 1.165 \frac{S}{D}$$

1. Weight of ship $= 0.1815\ L \times B = 2.0446 \dfrac{S^2}{D^2}$
2. Weight of engines $= S$
3. Passengers $= 0.03\ L \times B = 0.3379 \dfrac{S^2}{D^2}$

$$2.3825 \frac{S^2}{D^2}$$

"The quantity of coal at the mean draft will be $N \times$ daily consumption. The daily consumption depending upon speed and N depending also on speed, this must be settled

taking Calcutta	— 11500 miles	P. Phillip — 12500
at $13\frac{1}{2}$ knots	35.5 days	38.58
14	34.22	37.19
15	31.944	34.72

"Taking the $13\frac{1}{2}$ knots and consequently the length of voyage being $35\frac{1}{2}$ days and assuming $9\frac{1}{4}$ days spare coals,

$$N \text{ becomes} = \frac{35\frac{1}{2}}{2} + 9\frac{1}{4} = 27 \text{ days}$$

Total weight therefore of coal at the time of mean draft

$$= N \times \text{daily consumption} = N \times 0.11\ S$$

Weight of ship and coal and cargo at mean draft

$$= 2.3825\frac{S^2}{D^2} + S + 0.11\ N\,S + C$$

$$= \text{displacement} = 0.18372\frac{S^2}{D}$$

$$\left(\frac{2.3825}{D^2} - \frac{0.18372}{D}\right)S^2 + (0.11\ N+1)S + C = 0$$

$$S = \frac{(0.11N+1)D^2}{0.36744D - 4.765} + \sqrt{\frac{CD^2}{0.18372D - 2.3825} + \theta^2}$$

$$(\theta = 0.36744D - 4.765)$$

"Assume $N = 27$, $D = 24$, $C = 3000$

$S = 564 + \sqrt{852871 + 318096} = 1646$

$\therefore S = 1650$

$L = 665$

$B = 80.27$

Weight of ship	9656 tons	
engines	1650	
stores and passengers	1596	
cargo	3000	
coals halfway	5008	(Daily consumption $185\frac{1}{2}$ tons × 27 days)
	20910	
Displacement $0.18372\frac{S^2}{D}$ =	20840	
$P = 3.5 \times S$ =	5785"	

The procedure of equating the weight of the ship and its contents (using empirical formulae to relate the component weights to ship dimensions and speed) to the displacement (using typical dimensional ratios and form coefficients) and thereby deducing an equation in one of the design variables, remains today as a basic tool of the naval architect for preliminary design purposes. Although some of the relationships assumed by Brunel necessarily reflect the contemporary state of knowledge and the shortage of relevant data, it is in the way they are used to point inexorably to the hitherto unprecedented dimensions required that the passage quoted is of particular interest. The *Adelaide*, incidentally, was a mail ship designed and built by John Scott Russell for the Australian Mail Company. Brunel had advised the Company to build ships for the Australia run which needed to take on coal only at the South African Cape, and proposed accordingly that ships of 5,000 or 6,000 tons displacement would be required.

Examination of Brunel's calculations quoted shows how the question of whether coals should be carried for the whole round trip to the Orient or only for each half of the voyage was exercising his mind. The words "minus the halftime of the voyage" were clearly added in a second round of calculations. The decision to provide enough bunker space (over 12,000 tons) to fuel the round trip of the *Great Eastern* resulted in her displacement of 27,380 tons with length and breadth (680 feet by 82.5 feet) only slightly greater than those already deduced, and a draft fully laden of 30 feet.

Brunel's original thinking on the question of ship size found its first expression in the *Great Western*. Of the detailed naval architectural methods used in developing the design of that ship, little evidence remains. Wooden-hulled, copper-sheathed, with four-masted schooner rig, her performance in service, in relation both to speed and handling qualities, testified to the soundness of the basic design, much of which is doubtless attributable to Patterson, in whose Bristol shipyard she was built. In outward appearance she would have been remarkable mainly for her size. Internally, the architecture of the *Great Western* included one feature which was later—much later—to become an accepted part of the philosophy of ship design. For in his concern for the correct disposition and design of the structure of his ships, Brunel came to make one of his most important contributions to naval architecture.

Two quotations from one of his letter books *(PLB, 26 Feb., 1854)* illustrate well the attitudes he brought to bear on this subject:

"No materials shall be employed on any part except at the place and in the direction and in the proportion in which it is required and can be usefully applied for the strength of the ship—and none merely for the purpose of facilitating the framing and first construction", and "We should get on much quicker if we had no previous habits and prejudices on the subject". Another similar comment *(PLB, 10 Apr., 1855)* likewise informed his attitude, not only towards his structural work: "The most useful and valuable experience is that derived from failures and not from successes".

In formulating the design of the *Great Western* Brunel intuitively recognised that the transition from small to large ships brought increasingly into prominence the need for beam-like longitudinal strength of a ship's hull. The structures of very small ocean-going ships are determined largely by considerations of their ability to resist the lateral pressures of the sea which tend to distort the cross-section of the hull. The reasons for this have to do with the small probability of encountering those waves which cause significant longitudinal bending of the hull under weight and buoyancy forces. Above a length of around 150 feet these latter effects assume greater importance, so that for typical ocean-going merchant ships of today, with lengths in the range 300–900 feet, the provision of adequate bending and shearing strength in the hull, regarded as a hollow box beam, is a cardinal point of design.

The structural designs of bridges and ships have many points in common, and no doubt Brunel's experience in bridge work influenced the design of the structure of the *Great Western*. Certainly he was at great pains to ensure adequacy, and especially continuity, of her longitudinal strength, as he reported to the Great Western Steamship Company:

> "Her floors are of great length and over-run each other; they are firmly dowelled and bolted, first in pairs, then together, by means of 1½″ bolts, about 24′ in length driven in parallel rows, scarfing about 4′ . . . She is most firmly and closely trussed with iron and wood diagonals and shelf-pieces, which, with the whole of her upper works, are fastened with screws and nuts, to a much greater extent than has hitherto been put to practice" (Farr, 1963).

The integrity of the structure was amply demonstrated in the subsequent performance of the ship. Despite a stranding and twenty years of arduous service, there was no evidence of any defect in the strength of this ship.

But it was in the design of the structure of the *Great Britain* that the most important advance was made in this aspect of naval architecture. Brunel's familiarity with iron in civil engineering and railway works, and the evident technical and commercial success with which the material was beginning to be applied to the construction of small ships, encouraged him to recommend to the Directors of the steamship Company the use of iron for the new ship, although this was to be nearly four times the size of any iron ship yet built. These recommendations were evidently strongly influenced by the observations which Brunel and his associates had deliberately made during a voyage to Antwerp on the little paddle-steamer *Rainbow*, built of iron in 1837 by John Laird at Birkenhead.

What is not so certain is whether Brunel had by then (1838) heard of another small ship—the *Storm*—designed and built in 1834 by Scott Russell. Not only was she built of iron, but she was framed entirely longitudinally and fitted with closely spaced transverse bulkheads. The complete absence of transverse framing marks out this 70 foot vessel as probably the first radical departure from the transverse framing system hitherto universally applied in wooden ships.

But to Brunel must go the credit for first combining, on a dramatically large scale, the use of iron and of intuitive principles of ship strength. The resulting half-section of the *Great Britain* is shown in Fig. 34. Noteworthy features include the ten continuous longitudinal girders running the whole length of the double bottom, itself an innovation; the heavy ($\frac{7}{8}$ inch) keel plate which was welded into 50–60 foot lengths; the massive fabricated stem bar; the lapped double-riveted seams of the shell plating; docking keel bars external to the hull below the turn of bilge; the continuous iron cargo deck; extensive use of pillaring and some longitudinal bulkheads; and diagonal ties connecting deck and side framing to confer additional strength against racking forces. The structure weight of the ship was about 28 per cent of her displacement—a proportion which compares remarkably well with that for more modern cargo ships built of higher strength materials. A contributory factor to this high level of structural efficiency was the notion, first introduced in the *Great Britain*, that the scantlings of the structure could safely be reduced towards the ends of the ship where the external bending and shearing actions are less than near amidships.

Nor was there any evidence of inadequacy of her strength. Indeed few novel structures have had their integrity put to so stern a test. Stranded for nearly a year in Dundrum Bay, and subjected to over a

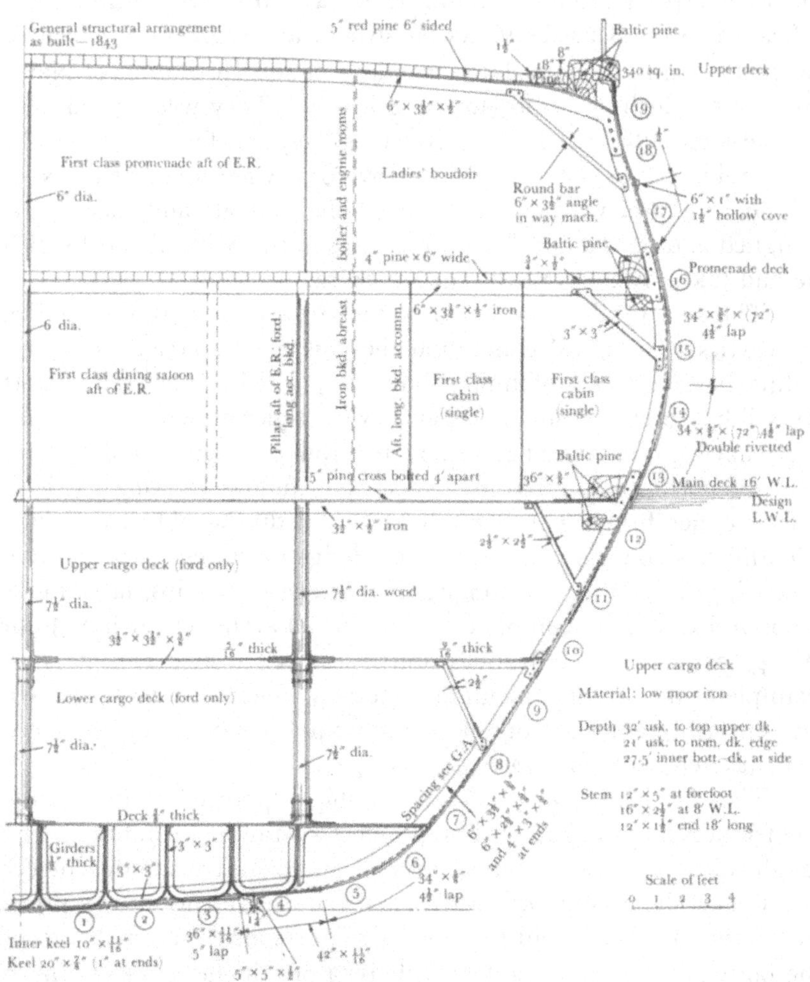

Fig. 34. Great Britain: hull section

century of various indignities (Corlett, 1971), this remarkable ship survived without serious structural trouble.

The coming together of Brunel and Scott Russell in the creation of the *Great Eastern* ensured, in respect of structural design (if not in all other aspects of naval architecture), a ship of truly historic significance. Scott Russell's views on structural design were presented at great length at many later meetings of the Institution of Naval Architects which he helped to found in 1860. They were as sound as his ideas on hull form design were erroneous, and for neither was he at all reluctant to claim credit. But by 1851, when the first ideas for the *Great Eastern* were being sketched, he had already successfully designed and built a number of iron ships in the Millwall yard which he had taken over from William Fairbairn.

The structural cross-section of the *Great Eastern*, shown in Fig. 35, carries to a logical conclusion the concepts of design for longitudinal strength and of hull subdivision, which Brunel and Scott Russell had been developing separately in earlier designs. It exemplified what Elgar expressed 40 years later (1894)—"optimum design (of ships) depends on personal judgement, hopefulness and imagination"—for neither Brunel nor Scott Russell had the benefit of any firm scientific basis on which to found their design decisions. The primitive beginnings of a theory of longitudinal strength of ships, later to be propounded in more complete form by naval architects such as John (1874), can be seen in Brunel's notebooks, from which Fig. 36 is an example of the kind of calculation method which he used to deduce appropriate scantlings for the longitudinal structure of his large ships (sketch-book *SB, steamship, 1852–54*).

The sketch in Fig. 36 shows the ship supported at two points 400 feet apart; this apparently represents a stranding condition. The rough calculations below reflect Brunel's uncertainty as to the correct positions of these "support points" and the equivalence of this problem to that of a beam, but the resulting figures for the cross-section of the bottom of the ship—90 feet wide by 4 or 5 inches thick—seem to cause him no concern. Possibly Scott Russell was able to persuade him (as he later argued before his fellow naval architects) that the stranding condition was intolerably severe as a design loading case. It would seem that in the eventual design of the *Great Eastern* structure, their views, on this aspect of naval architecture at least, were entirely in harmony.

Brunel's own notebooks show he was much concerned in the early 1850s with the problems of ship strength and seaworthiness. In

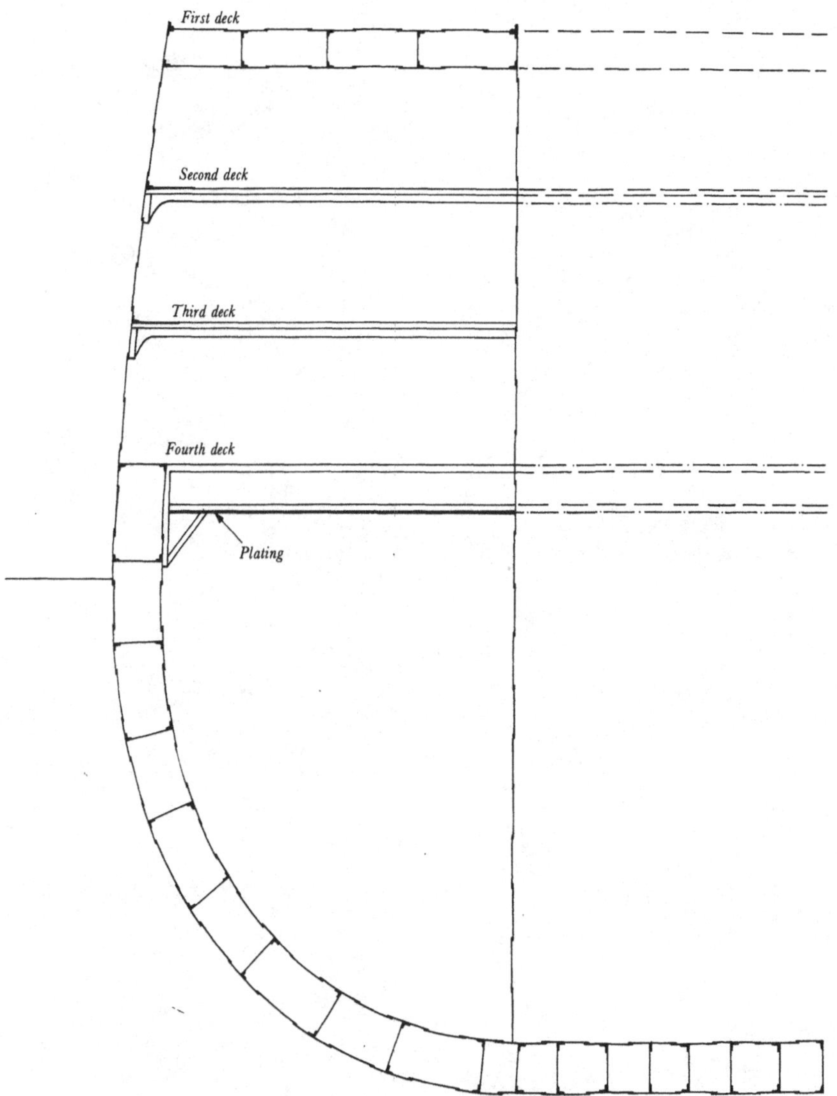

Fig. 35. Great Eastern: hull section

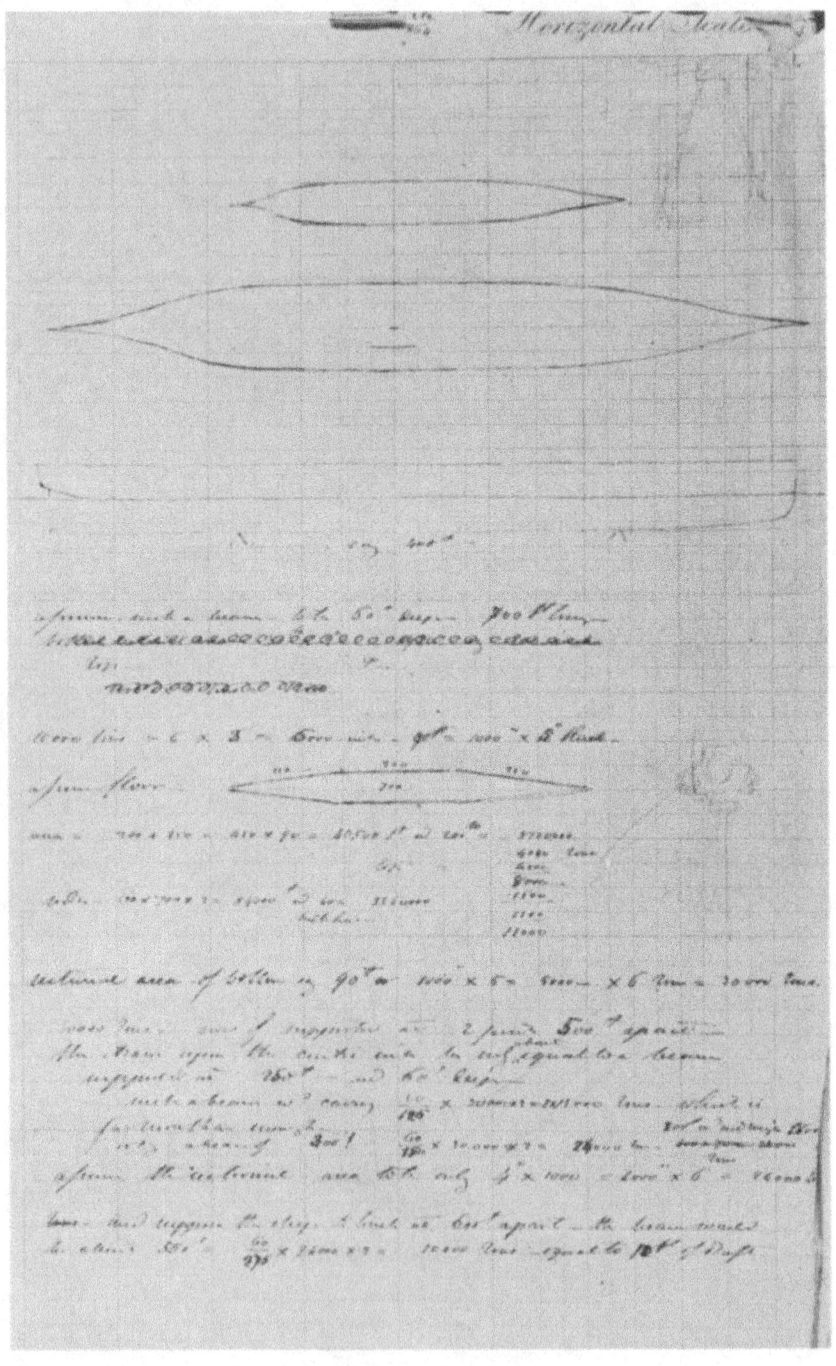

Fig. 36. Analysis for longitudinal structure

1854 he writes of the *Great Eastern*:

"We have flattened the floor and strengthened it considerably so as to allow of the vessel being safely grounded on a gridiron or even if partially waterborne on a beach. The parts which seem to me now to admit most advantageously of being strengthened are the main deck, and the sides connecting these with the bottom.

"First as regards the iron deck or top web, there appears to be about 620 sq ins only equivalent say to $3500 \times 2 = 7000$ tons of strain. This is but small for any real strength of the ship as a beam—and yet it weighs upwards of 900 tons—so that unless one adds some 200 or 300 tons to it there is not much to be gained. However I suppose we may assume that it would bear nearly 10, say 8, tons before breaking, which gives $620 \times 2 \times 8 = 9920$ tons of tension and assuming the beam to be 55 ft deep the ship would almost carry herself supported on two points at 300 ft apart. To bring them into full operation however, the vertical web or top sides must be stronger, and I think if 150 feet runs at about the $\frac{1}{4}$ and $\frac{3}{4}$ of the length were very much strengthened, it would be a good thing. This would involve, of each $\frac{1}{8}''$ of thickness, —150 ft long × 22 ft high × 4 no × 5 lb = 66000 lb, say 30 tons. I think this would be worth applying and also—good material stiffening to prevent buckling." (Eastern Steam Navigation Company *ESN, 1854*)

Brunel's intuitive "feel" for structures has again identified two aspects of hull girder behaviour which were not fully explored until much later: the need for adequate shear stiffness in the side "webs" (especially near the quarter points of length) and the need to guard against plate buckling. Nor did Brunel fail to respect the lessons of experience about the forces of the sea. In some almost indecipherable notes *(ESN, 1854)* which must have been made in the late 1850s, he writes:

"After a careful examination of the *Persia*,[37] and thinking a good deal on the requirements of sea-going steamers, and recalling various experiences—the performance of the Great Britain and her 'thumps' in the Bristol Channel amongst other things . . . I come to the following conclusions (for a very large and high ship) . . . a small flare out—*not quite* upright, . . . bulwarks . . . enough to be certainly above the tops of a breaking sea; . . . there should be a forecastle deck of some length; . . . no water should be allowed to break *on* to a deck unless it can sweep off clear. Everything must be so found that if a solid sea swept from the bows right along the sides . . . it should meet with nothing flat to break fair against, such should be slanted off outwards."

The structure of the *Great Eastern* proved more than equal to the demands of 30 years of varied service, including a severe explosion and a grounding, and then took more than three years to break up. Her midship section developed and extended the concepts introduced in the *Great Britain*. The remarkable double-skinned upper deck structure ensured at last an effective upper "flange" to the hull girder where previously most designs had wrongly relied on a wooden upper deck spanning the gap between the side frames. Further massive girder strength is provided by the ½ inch thick longitudinal bulkheads and by the double hull structure, itself well stiffened longitudinally, extending beyond the bilge to the lower deck. The merits of this latter feature were not only structural. When the ship later ran on to a reef which breached the outer hull over a length of 80 feet, the presence of the inner shell—then a quite new feature in ship construction—undoubtedly saved a major disaster. It was reported that, in the event, the passengers were unaware that anything untoward had occurred.

Not the least extraordinary feature of this design, and one which even today's shipbuilders might do well to emulate, is the degree of standardisation of scantlings which Brunel achieved. Standard plate sizes and thicknesses (1 inch, ¾ inch and ½ inch), bar sizes (4 inch by 4 inch by ⅝ inch), rivet diameters (⅞ inch) and arrangements (single-riveted seams, double-riveted butts, 3 inch pitch) were used throughout. No ship of comparable size has since been built of so few standard elements. Despite the increase in structural weight that such standardisation generally implies, the ratio of hull structure weight to deep displacement of the *Great Eastern* was rather less than a quarter. Foster King's description 50 years later (1907) of her structure as "a monument of successful design, whose lessons have not yet been applied in practice" still stands today.

Other naval architectural features of the *Great Eastern* were, however, not so successful. The troubled story of her launch, sideways into the River Thames and taking 90 days to complete, is well known. Brunel's successful experience of the "floating out" of the *Great Britain* (which, however, was not without its problems) no doubt inclined him to prefer to construct the giant ship in a dock, but the cost and civil engineering problems caused this original scheme to be abandoned. Committed thereafter to a side launch, Brunel devoted great care and thoroughness to ensuring its success; but despite his many experiments, calculations and correspondence with American protagonists of side launching, the uncertainties of frictional effects in his metal to metal launchways and of the behaviour of their piled foundations resulted in a painfully slow and costly launch.

As to the sea-going qualities of the *Great Eastern*, the later evidence is conflicting. Brunel himself seems to have taken a rather lesser interest in those aspects of naval architecture which concern hull form design, sea-keeping and ship motions, although the mechanics of steering and the design of rudders gave him considerable cause for thought and development of ideas. The hull forms of his first two ships were mainly due to Patterson, and that of the *Great Eastern* resulted from Scott Russell's notions of ship resistance, as embodied in his "wave-line" principle (1860, 1861), on which he had been working since 1834. That she did not quite achieve the speed performance expected of her is, in retrospect, unsurprising in view of the state of the art of resistance prediction coupled with the basic fallacies inherent in Scott Russell's ideas. It was to be another fifteen years before the genius of William Froude began to unravel the complex dependencies of ship resistance on hull form, which laid the foundations of modern methods of power estimation.

But Froude was well known to Brunel as a contemporary and worked for him for a time (see Chapter I) as a civil engineer on the Bristol and Exeter Railway. Even if it is doubtful whether Froude influenced significantly Brunel's ideas on ship design, there can be no doubt that the problems posed by Brunel in designing the *Great Eastern* greatly influenced the direction of Froude's own work. Indeed it may well be that naval architecture owes to Brunel a most historic debt: the engaging of Froude's interest in problems of ship dynamics.

Certainly the theoretical work on ship rolling which Froude presented to the Institution of Naval Architects (1861) had been stimulated in part by tests on the *Great Eastern*. The ship evidently suffered from excessive rolling, the paddles frequently emerging from the water. Her designers had no more than past experience and simple notions of statical stability on which to judge the likely rolling behaviour of the ship in a seaway. In the event, the combination of very large metacentric height and very large bilge radius (without bilge keels) led to a ship whose natural period and low damping resulted in a more frequent synchronism with deep sea waves, and with consequently large roll angles. From his investigations Froude was able to construct the first satisfactory analysis of the dynamic stability of ships, which remained essentially the accepted basis of ship dynamics for the next 100 years.

The manoeuvring qualities of the *Great Eastern* showed contrary aspects. On the one hand the combination of paddles and a screw offered considerable versatility, so that very close control of forward and rotational motion could be exercised. Perhaps this was a fortunate consequence not foreseen by her designers (these qualities proving especially useful in her service as a cable layer)—the reason for combining screw and paddle propulsion was that it was believed that the designed power could not be obtained from screws alone. The original steering arrangements for normal steaming operations, however, proved unsatisfactory. Despite a brief note in one of Brunel's calculation books—"to have very powerful steering apparatus—with mechanical assistance if it can be easily managed" *(ESN, 1854)*—the giant ship was originally steered by hand, and proved exceedingly difficult to control in this way. Subsequently Gray (1869) implemented Brunel's concept by fitting power-operated gear to the *Great Eastern*'s rudder. Brunel's own recognition of the problems of controlling the rudder by manpower show clearly in his calculations *(ESN, 1854)*. After commenting on the evident inadequacy of the steering apparatus on the *Persia*, he deduces that with the greater rudder size and proposed speed for the new ship the rudder resistance would be eleven times that on the *Persia*:

> "Now 4 men at the wheel of 6 ft diam. requiring $2\frac{1}{4}$ turns wd. be totally unsufficient—say 8 men working through a space of $2\frac{1}{4} \times 6$ = say 42.75 ft. And for Eastern Steam, $8 \times 11 \times 42.75 = 1$ man moving 3762 ft. Say 30 lb per man. 3762×30 = say 50 tons $\times$ 1 ft. But this seems to be a very small force to apply to the rudder to resist the motion of the rudder. What are the forces?"

Then by finding the surface pressure on the rudder to be equivalent to a head of about 10 feet and assuming a rudder angle of 40 degrees, the effect on the rudder is found to be equivalent to a force of 150 tons at 1 foot leverage:

> "This seems enormous—and yet quite possible—at all events one can quite conceive a wave combined with the motion of the ship producing a head of 5 ft directly at right angles against the surface of the rudder . . . Now this would produce a direct strain equal to 224 (rudder area) $\times 5 \times 64 \times 32$ tons acting with a lever of 5.5 = say, 175 tons!! It seems therefore that provision should be made to resist more than that strain . . ."

Clearly the problem of controlling the *Great Eastern*'s rudder was going to call for considerable human effort. A surprising feature of this aspect of the design is that Brunel seemed to have abandoned his belief in the "balanced" rudder, which he fitted earlier in the *Great Britain*. A side elevation of the original stern arrangement of the latter ship is shown in Fig. 37. As Corlett has remarked (1971), "it is doubtful if any ship within the next fifty years was fitted with a more

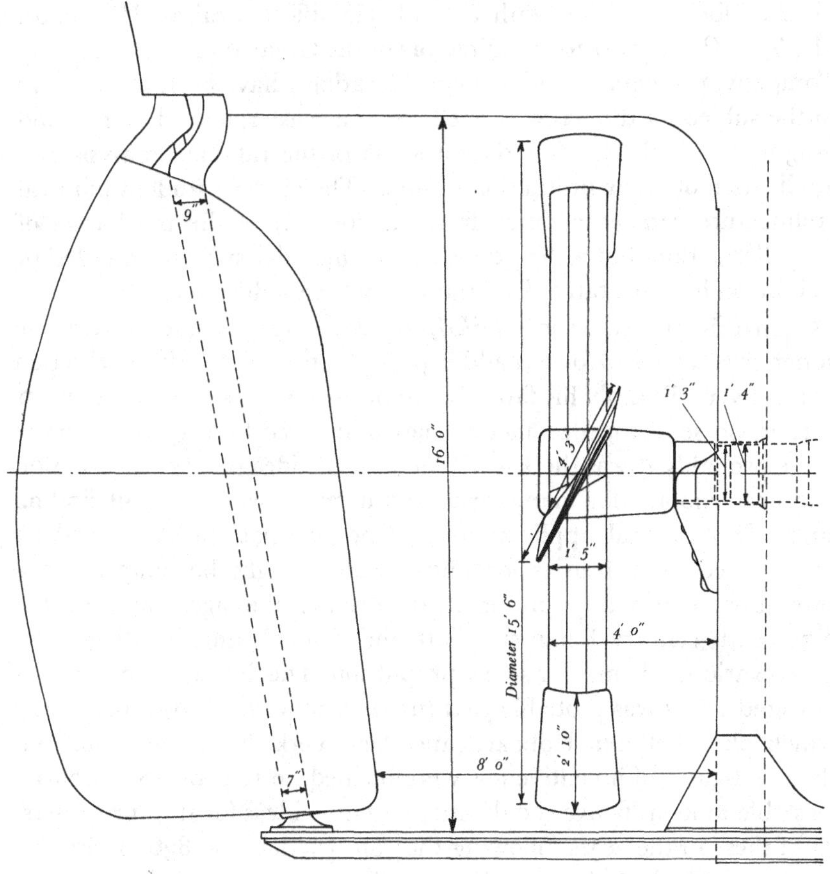

Fig. 37. Great Britain: stern arrangement

modern and efficient arrangement". Although the origin of this type of rudder probably goes back to Earl Stanhope's "equipollent rudder" patented in 1790, Brunel characteristically recognised its merits in reducing the power required to operate the rudder by having part of the rudder forward of the hinge line, and accordingly boldly adopted it for his new ship. No doubt it was the rudder which was subsequently blamed for the reportedly poor handling qualities of the *Great Britain.* Certainly controversy as to its real merits continued for many years, and probably influenced the decision to revert to a more conventional rudder design for the *Great Eastern.*

But on one feature of the *Great Britain* there was no dissent. It is for her pioneering role in screw propulsion that she has been rightly acclaimed. In this development we see Brunel combining, at his best, his flair for innovation with careful scientific appraisal. His report *(PLB, 10 Oct., 1840)* to the Directors of the Great Western Steamship Company, presenting "the best consideration I have been able to give to the subject of the screw propeller", is a masterpiece of clarity and insight; it is indeed a foundation stone of the rational analysis and application of screw propulsion of ships. The modern student of naval architecture can learn much from it, not only of the mechanics of propeller action, but also of the way in which decisions in ship design call for both quantitative fact and qualitative judgement.

In this historic report *(PLB, 10 Oct., 1840)* Brunel argues the superiority of screw over paddle propulsion with all the conviction of a convert. Possibly his father's exploratory tests, some twenty years earlier, on self-propelled models had influenced him against screws; more probably it was the knowledge that the idea of screw propulsion had been mooted for many years and in many ways without finding successful practical application. The notion that the Archimedean screw, used for centuries for lifting water, might be adapted as a propulsor, had many protagonists through the ages, and in the eighteenth century Bouguer, Watt and Bramah among others proposed various schemes for screw propulsion. The first sea-going steam-powered screw was probably that fitted by Stevens in 1804, on a ferry which plied between Hoboken and New York. From this point on, the flood-gates of inventiveness were opened, so that by 1852 Bourne was able to identify over 90 distinct patented ideas for ship propellers, 74 of these in the years following the construction, in 1836, of Smith's screw-propelled *Archimedes*. Nearly all those variants from single, fixed-pitch propellers which are in vogue or under investigation today were anticipated among this extraordinary array of patents: contra-rotation, variable pitch, overlapping twin screws, ducted screws, even screws with flexible blades, and such exotic notions as the "boomerang propeller" and the "bird's wing blade". Bourne himself contributed some novel ideas in relation to ship propulsion, ranging from a confident denunciation of Newton's laws of motion to a rather more defensible scheme for reducing ship frictional resistance by admitting air along the hull below the water-line.

Brunel's distaste for patent procedure may have further prejudiced him against such ideas on screw propulsion; practical possibility clearly mattered more to him than patented fancy. Perhaps the real potential of screw propulsion was first borne in on him when his own *Great Western* was towed at 2½ knots into the East India Dock by a little 34 foot long screw-propelled boat built by Smith. To promote and develop his ideas, Smith later built the *Archimedes* (125 feet long, 21.8 foot beam, 9.5 foot draft), a 237 ton vessel developing 90 horsepower from her twin-cylinder engine driving a 5.75 foot diameter screw of 8 foot pitch. It was on a "sales tour" in 1840 for Smith's Screw Propeller Company that the *Archimedes* came to Brunel's notice, mainly through the good reports of her performance which he received from Thomas Guppy, then a leading member of the building committee of the Great Western Steamship Company.

Construction of the *Great Britain*, designed for propulsion by steam-powered paddles and by sails, was at that time well under way. Brunel advised that work on the paddle engines be suspended while he considered further the matter of screw propulsion. That there followed such a radical change in the construction of a ship whose design was already exceptional in size, material, machinery and layout testifies to the completeness of Brunel's conversion, and to the invincible arguments for screw propulsion propounded in his report.

He first discusses "slip"[38] as one indicator of efficiency of a propulsor in converting engine power to useful thrust. Recognising the impossibility of calculating slip, he shows by experiments that the average slip in *Archimedes* of 29 per cent compares very favourably with the 27 per cent of the *Great Western*, despite the fact that the area of the propelling surface in the *Archimedes* was only 20 per cent of its midship section area, compared with 40 per cent in the paddle boards of the *Great Western*. This:

> ". . . may at first appear startling . . . but having witnessed and carefully observed . . . the nature of the disturbance in the water . . . I no longer feel surprised . . . it is generally assumed that the inclined plane formed by the thread of the screw strikes the particles of water at that angle and with the velocity of the revolutions of the screws, but it is forgotten that the screw is moving forward with the ship . . . ."

This crucial deduction and Brunel's subsequent reasoning represent the beginnings of a correct interpretation of the action of a screw propeller, later to be developed by Rankine, Froude and others into the complex theoretical analyses in use today.

The adverse effect of friction is next considered, and again observation followed by deductive reasoning leads to insight. The absence of any apparent rotary motion of the water in the wake leads Brunel to suppose that, although frictional forces will inevitably diminish propeller efficiency and will be exacerbated by blade roughness, these effects are likely to be small. But Brunel disclaims a complete understanding of the action of the screw in developing thrust, not surprisingly perhaps, since the concept of a lifting surface had not yet been discovered. Missing therefore from Brunel's appreciation of propeller action is the lifting effect of camber of the blade surface. But he says that "the facts as proved by the experiments are what I rely on" and goes on to use his facts to deduce power requirements for screw-propelled ships. Both *Archimedes* and *Great Western* are shown to require about 144 horsepower per square foot of midship section, despite the smaller area of propelling surfaces and the evidently very rough surface of the screw in the *Archimedes*. Ways of improving its efficiency are then considered; slight increases in diameter and pitch would result in a required power reduction of one third.

In the second part of his report Brunel turns from the mechanical action of the screw to other more general aspects of its use. In so doing he anticipates nearly all those problems and difficulties which have subsequently attended screw propulsion: the need for correct stern forms ("a clean run is the most essential condition") and adequate propeller immersion, liability to damage, and the matching of propeller and machinery to ensure smooth and vibration-free running. Against each objection, observations and deductions are marshalled to show how—at least in the *Great Britain*—all such problems can be overcome. Finally a number of positive advantages are listed and analysed: substantial saving in weight, improvement in hull form and space within, ready adaptation of screw or sails to the prevailing weather conditions, absence of shock loads on engines, improved steering through the action of the propeller race on the rudder and, by no means least, a substantial reduction in the beam of the ship through the elimination of paddle-boxes.

The impact of this brilliant exercise of experiment, observation and reason could hardly fail to persuade the Directors of the "wisdom, I may almost say the necessity of . . . adopting the improvement" recommended by Brunel. History does not record whether eyebrows were raised at his relegation of cost considerations to a rather superficial paragraph of only four lines in an essay having roughly the length of the present chapter. But history amply vindicated his judgement, both of the general concept, and in some important particulars. Viewed through contemporary eyes the six-bladed propeller originally fitted on the *Great Britain* (15.5 foot diameter, 25 foot pitch, weighing nearly 4 tons) was not without defects: it was rather too large in diameter, located too near the stern-post (see Fig. 37) and of somewhat inefficient construction to withstand the forces involved. That subsequently there were failures of the propellers fitted to the *Great Britain* is not, therefore, surprising. But recent tests on a model of this propeller show that its open-water[39] peak efficiency (around 70 per cent at its best operating conditions) is only about 8 per cent less than that obtainable from contemporary designs. Brunel's estimate that *Great Britain* would make 12 knots at 54 revolutions per minute was astonishingly good: on trials 12⅓ knots were achieved at 55 revolutions per minute. Furthermore, this rate of revolution has recently been shown (Sinclair, 1971) to be almost ideal for the propeller!

Brunel's later work on propellers substantiated, rather than extended, his understanding, and provided further data for design. His wrangles with the Navy, who commissioned further experimental work by him, are well recorded by Rolt (1957). The trials of *Rattler* and *Alecto*, further confirming the superiority of the screw, are a part of the history of naval architecture. In this aspect of ship design, art and science were now being joined in fruitful partnership.

But in finally assessing Brunel's work, the naval architect's admiration will generally fall short of idolatry. The successful creation of any new design of ship calls for the exercise of many talents and particular skills never to be found in one man, even a Brunel. It was Brunel's good fortune that men of the abilities of William Patterson, Thomas Guppy, Christopher Claxton and John Scott Russell were able to catalyse his genius, and thereby to give practical effect to his novel ideas which rapid changes in technology were then making possible. And if it is difficult to apportion credit for these engineering achievements, we should perhaps refrain also from allocating blame for the subsequent failure of the three ships to achieve the commercial success for which their progenitors had hoped. Writing of them some years later, John (1887) took a stern view: "the highest flights of constructive genius may prove abortive if not strictly subordinated to the practical conditions and commercial requirements of the times". He was thinking of the three Brunel ships, and how the *Great Western* and *Great Britain* had failed to capture the transatlantic trade for Bristol, not through technical inadequacy but because the circumstances of their operation, and especially the port facilities at Bristol, were not adapted to the new trades and capabilities of the ships.

Brunel's later attempts at techno-economic calculations for the *Great Eastern* helped to convince the Directors of the Eastern Steam Navigation Company that the ship should earn a 40 per cent per annum return on the capital invested. That the ship was simply too big for the volume of cargo or numbers of passengers then using the routes for which she was designed appeared to escape their considerations. Brunel himself was evidently concerned about this. In a conversation with Robert Stephenson and W. S. Lindsay (1876) at Millwall, where the *Great Eastern* was nearing completion in the summer of 1857, he asked his visitors: .

> "How will she pay? If she belonged to you, in what trade would you place her?"

Lindsay's forthright reply upset him:

> "Send her to Brighton, dig out a hole in the beach and bed her stern in it, and if well set she would make a substantial *pier*, and her deck a splendid promenade; her hold would make magnificent salt water baths and her 'tween decks a grand hotel, with restaurants, smoking and dancing saloons, and I know not what all. For I do not know any other trade, at present, in which she will be likely to pay so well."

Nor indeed did the *Great Eastern* pay well, or even operate, on the Far East run for which she was designed, nor on the Atlantic immigrant trade where her great size might have been better utilised. What commercial success she did achieve was as a cable layer. Her failure was not that she lacked the "fitness for purpose" which is the essence of successful engineering, but that her purpose was wrongly identified. Was the engineer then to blame?

Perhaps to Brunel the acts of creation and of innovation were more important than their eventual commercial consequences. Each of his three great ships, as we have seen, advanced in some major and lasting way the science and practice of naval architecture. Much more could be written of his contributions to marine engineering, of his ideas and sketches on rigging, navigational aids, experiments on materials, on launchway friction, jet propulsion, lifeboats, his plans for a floating siege gun . . . But it was his successful demonstration that new things *could* be done, and especially that scientific and numerate argument could help to create and justify such innovations, which gave naval architecture new excitement and new confidence.

A year after Brunel's death in 1859 the Institution of Naval Architects was founded. Its *Transactions* record the subsequent development of those ideas for which these three ships are justly famous. More than that, they record the eventual coming together of the traditional art and the emerging science of naval architecture. In the golden years which were to follow, these legacies of Brunel's work ensured for him a special place in the history of ships and the sea.

VIII

Royal Albert Bridge, Saltash

SIR HUBERT SHIRLEY-SMITH

In 1845 the Cornwall Railway Company applied for an Act to extend the railways westwards from Plymouth to Falmouth. This involved the crossing of the River Tamar either by means of a steam ferry at Torpoint or a bridge at Saltash. Parliament sanctioned the bridge and so was commissioned Brunel's last and greatest bridge building achievement.

The site at Saltash, about three miles north-west of Plymouth, was chosen because there the river narrows to a width of 1,100 feet with rock outcrops on either side and has a depth of about 70 feet above the mud. Brunel, who was the Engineer for the Cornwall Railway, could foresee the high cost of the deep river piers of a multi-span bridge and seriously contemplated even bridging the river with a single span. He felt quite competent to build a span of 1,000 feet to carry a line of railway traffic and had indeed calculated that the total cost of such a bridge in wrought iron would be about £500,000.

Early in 1847 a preliminary examination by means of borings was made of the bed of the Tamar river. This showed that on rather more than half of the span on the eastern side there was rock overlain with from 3 feet to 16 feet of mud, but that to the west the level of the rock fell away rapidly until the greenstone trap outcrop of the Saltash cliff was encountered.

Brunel ordered a thorough examination of that part of the river, about 50 feet square, where the centre pier of a two-span bridge would be located. This was carried out by means of a cylinder of wrought iron, 85 feet long and 6 feet in diameter, which was slung from two gun-brig hulks and lowered vertically to the bed of the river. It is an example of Brunel's thoroughness that no fewer than 175 borings were made inside this cylinder at 35 different places where it was pitched. Ultimately the cylinder was sunk to the rock, the water pumped out, the mud excavated inside and trial masonry built on the rock up to river bed level before the cylinder was withdrawn.

After careful consideration of all the factors, Brunel decided to use two wrought iron spans of 465 feet each and only one river pier. Two years later, under financial stringency, a single line of railway was substituted for the double track, thus saving a cost of £100,000, and it was considered practicable to reduce the spans to 455 feet each, which Brunel had no compunction in doing. This meant that the spans would be 5 feet less in length than those of the Britannia tubular bridge recently completed by Robert Stephenson—but such petty-minded considerations would never have been entertained by Brunel. The restriction to single line trains does not cause inconvenience at Saltash because performance is so greatly improved by modern methods of signalling, in spite of the ever-increasing traffic of today.

The final design (Fig. 38), on which work began at site early in 1853, was for a single track railway bridge with an overall length,

Fig. 38. Royal Albert Bridge as completed British Rail

including curved approaches, of nearly 2,200 feet. The bridge comprised two spans of 455 feet centre to centre of piers (with a minimum 100 feet of headroom at high spring tide), two spans of 93 feet, two of 83 feet 6 inches, two of 78 feet, two of 72 feet 6 inches and nine of 69 feet 6 inches. Such small variations in the length of the approach spans would not be acceptable or economic in the fabrication shops of today, but standardisation and repetition were obviously of much less importance 120 years ago than they are now.

Each of the two great trusses over the river, by which the bridge is so well known, consists of a wrought iron oval tube in the form of a parabolic arch, with the ends tied by two pairs of suspension chains, one on each side of the tube. The distance between the eye attachments of the suspension chains is 450 feet, the total length of each tube being 461 feet. The rise of the arch is equal to the dip of the chains, the overall depth of the truss being 56 feet. The tubes of the two trusses are joined over the centre pier where the bearings are fixed and there are expansion bearings on $3\frac{1}{2}$ inch diameter wrought iron rollers at the outer ends. Each span is divided into twelve panels by means of eleven vertical members braced with ten pairs of diagonals. The two main longitudinal girders carrying the railway deck are suspended from the trusses by hangers beneath the verticals and intermediately. The total weight of each main span, including trusses and deck, is about 1,060 tons.

The main spans of the Saltash bridge are composite prestressed structures in which the pull of the suspension cables resists the thrust of the arch overhead. It can thus be considered a self-anchored suspension bridge, or a tied arch or even a rudimentary form of truss. But if we accept the first of these definitions, then Brunel has the distinction of being the only British engineer to build a suspension bridge that has proved capable of carrying main line railway traffic successfully, in spite of the ever-increasing weight of the trains, and has continued to do so for more than a hundred years. And when we pause to think that, in this same decade, Robert Stephenson with the 460 foot long box girders of the Britannia bridge (in 1850) had built the forerunner of all the steel box girders of today, we realise what a wonderful era of achievement this was.

Let us look now at one of the greatest problems that faced Brunel at Saltash: the design and method of sinking of the big centre pier (Brereton, 1861). This was to consist of a circular column of solid masonry 35 feet in diameter and 96 feet high from its foundation on bedrock to its upper surface above high water level. On top of this were built four octagonal columns of cast iron, 100 feet in height, to the level of the railway, with cross-bracing of cast iron but no longitudinal bracing between them. The columns measured 10 feet across in section and had a wall thickness of 2 inches with internal stiffening. They were erected in lengths of 6 feet with flanges interconnected by vertical bolts. Above the top of the columns was located the centre portal, of similar construction to the octagonal columns. Some 50 feet in height, this portal served not only to cover in and roof over the ends of the big wrought iron tubes and chains but also to incorporate in its base the fixed centre bearings of both spans. On the two outer piers of the main spans masonry was used in place of the octagonal cast iron columns; the end portals are of brick with courses of stone, all encased in cast iron. It was estimated that the dead load reaction at the base of the centre pier would be 9½ tons per square foot, increased to about 10 tons per square foot by live and other loads—a pressure the rock was well able to withstand.

Brunel decided that a wrought iron cylinder of 35 foot diameter must be sunk, by means of compressed air if necessary, and that its base or cutting edge should be inclined, so that it was 6 feet deeper to the south-west, to suit the inclination of the rock surface. About 20 feet above the cutting edge a wrought iron dome was made to form the roof of the working chamber, and from the middle of this a 10 foot diameter shaft, open at the top and bottom, extended to the surface (Fig. 39).

Pneumatic sinking was a relatively new process, but it had been used successfully by Cubitt and John Wright in founding the 61 foot deep cylinders for the Rochester bridge over the Medway in 1851. Brunel had gained some experience of compressed air in diving bells on the Thames Tunnel and had used it on a small scale on the foundations of the Chepstow bridge. An interesting innovation of Brunel's at Saltash was the construction of an annular space 4 feet wide right round the circumference inside the working chamber. Brunel states in one of his letters (I. Brunel, 1870) that this idea was first suggested to him by R. P. Brereton, who was engaged under him on the design and construction of the bridge. The intention was to pump air into the annular ring only, thus expelling the water so that men could go in and make a cofferdam, as it were, at the bottom of the cylinder, without having to resort to the use of air pressure over its whole area.

The air jacket was further divided by iron partitions into eleven equal compartments and was connected at the top to a 6 foot diameter cylinder fixed eccentrically inside the 10 foot shaft. The 6 foot cylinder was used for the application of compressed air, as necessary, to the annular ring, and was connected to it by means of an air passage below the dome. Incidentally, Brunel achieved economy, in a manner followed by good designers since, in using permanent materials first for any temporary requirements. Thus the 6 foot diameter air shaft was the one previously used for making the borings and the annular ring was constructed of iron plates subsequently to be used in the deck girders of the bridge.

The big cylinder was built on the river bank, floated out, moored in position and lowered vertically to the river bed. The pneumatic apparatus used previously on the Chepstow bridge was brought into use, the air locks being assembled on top of the 6 foot air shaft. The air pumps were operated by two 10 horsepower steam engines on top of the 35 foot cylinder, and two 13 inch water pumps were installed inside the 10 foot cylinder. By means of water pressure on the dome and the use of iron ballast on the higher side, the cylinder was slowly forced down 3 feet into a nearly upright position. Men were then able to go down under compressed air through the 6 foot shaft and the air passage below the dome to excavate the compartments of the annular ring and hammer away the irregular rock surface. Two 2 foot diameter bucketways were provided for bringing out the muck.

The cutting edge of the cylinder was then 82 feet below high water level, corresponding to an air pressure for the men of about 35 pounds per square inch. By pumping out some of the water inside the 10 foot cylinder this pressure was reduced, but even so the 7 hour shifts worked proved far too long and many of the 30–40 men employed suffered from "the bends" or temporary paralysis. Very little was then known of the effects of working in compressed air, but when the shifts were reduced to 3 hours in length there was no more caisson sickness.

After much trouble with leaks through fissures in the rock the cylinder was sunk to its full depth of 87 feet 6 inches and a ring of granite ashlar 4 feet thick and 7 feet high was built in the air jacket. To provide against buoyancy in the event of compressed air having to be used over the whole area, 350 tons of pig-iron and kentledge were stacked on shelves in the upper part of the cylinder and another 400 tons placed on top of the dome. These measures were successful in weighting the cylinder down and when pumping was restarted the whole working chamber was emptied of water and men were able to go in to excavate the mud and level the rock in the open. Thus the early hopes of Brunel and Brereton were realised and once the

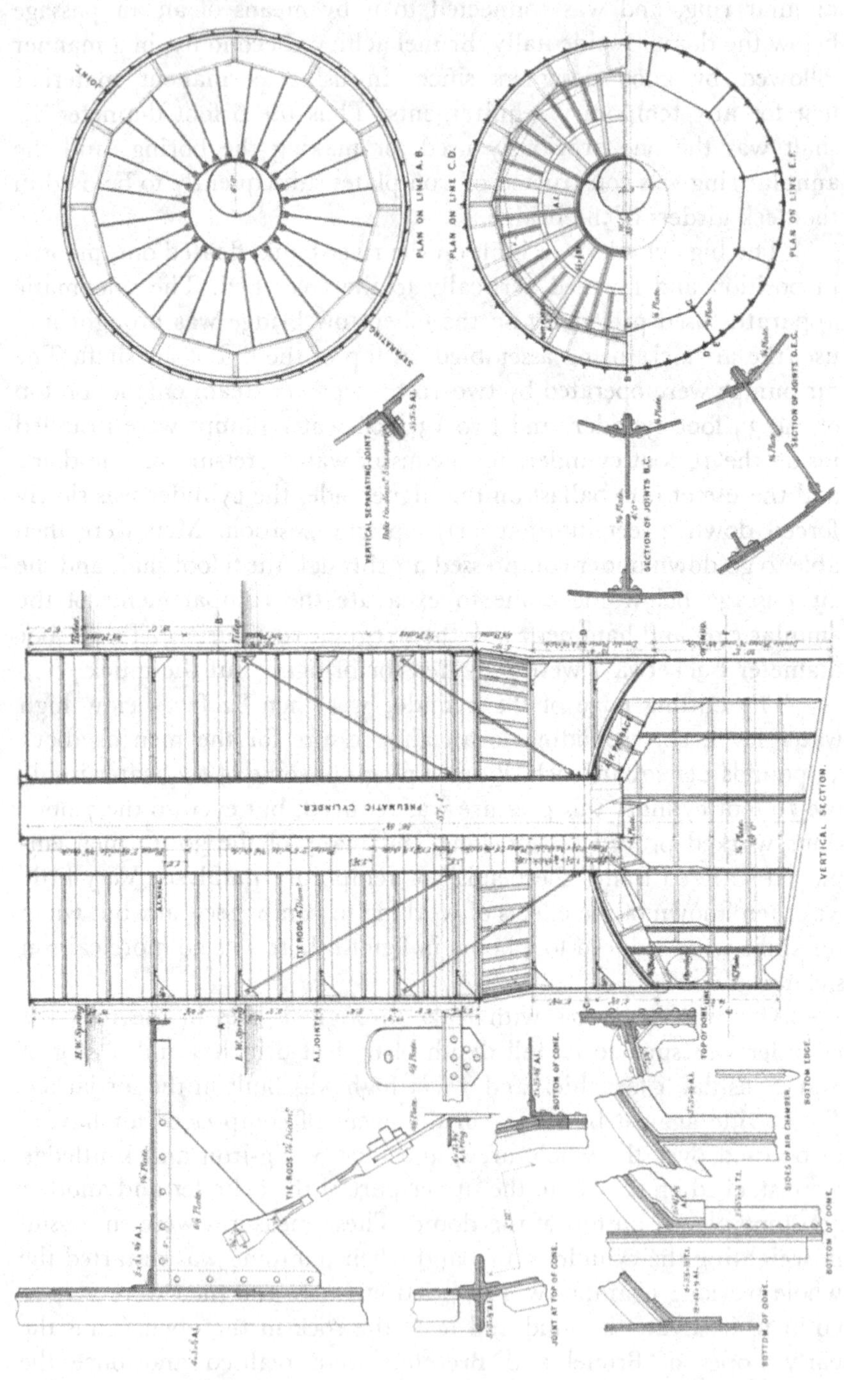
PLAN ON LINE A.B.
PLAN ON LINE C.D.
PLAN ON LINE E.F.
SEPARATING JOINT
VERTICAL SEPARATING JOINT
SECTION OF JOINTS B.C.
SECTION OF JOINTS D.E.F.
PNEUMATIC CYLINDER.
VERTICAL SECTION.
TIE RODS
TOP OF DOME
BOTTOM OF CONE.
BOTTOM EDGE.
SIDES OF AIR CHAMBER.
BOTTOM OF DOME.
JOINT AT TOP OF CONE.

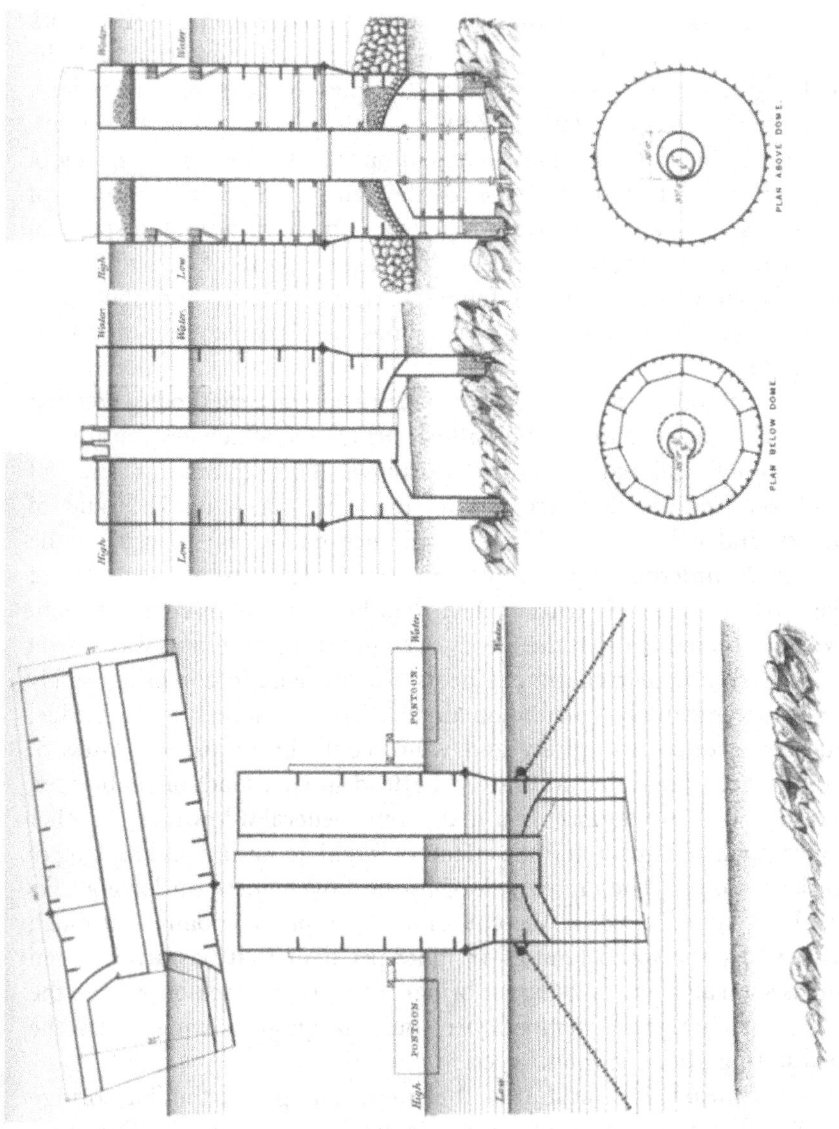

Fig. 39. Royal Albert Bridge: cylinder used for construction of centre pier

annular ring had been properly founded under compressed air it acted as a reasonably watertight cofferdam, enabling the whole of the centre part to be bottomed up in the open.

As the inner plates of the annulus were cut out, the new granite ashlar was bonded into the outer wall already built. So the work proceeded, with the removal of the temporary ironwork and the building up of the masonry, until the pier reached its full height, with the cap completed ready for the erection of the cast iron columns.

From the date of commencement on the cylinder in the spring of 1853 to the date of completion of the centre pier in the autumn of 1856 was 3½ years. This is a long time compared with the speed of construction achieved by the sophisticated plant of today, but it goes to show the degree of tenacity, determination and sheer grit needed to design and build a pier such as this—a prototype of its kind—120 years ago.

Let us look now at the design, fabrication and method of erection of the two main spans of the Saltash bridge for which few, if any, of the original calculations are in existence. Throughout his years of work on bridges, although Brunel made the best use he could of theory and calculations, his final reliance had to be placed on the results of numerous experiments and tests to destruction on models of the various forms of struts and girders he had under consideration. We must remember that the theory of engineering structures was not yet developed enough to enable engineers to design a large iron bridge safely and economically by calculation alone, even if the forces on the bridge, due for example to gravity and wind, could be accurately assessed. Brunel could determine accurately the dead and live loads on his bridges, but he could do no more than make some general allowance for other factors such as impact and the effects of lurching, nosing and rocking of the locomotives. The stresses caused by the movement of the locomotives would be aggravated by the curves of the approach spans, but much reduced by the speed limit of 15 miles per hour that has always been imposed on all trains crossing the bridge. Brunel could no doubt assess the effects of changes in temperature, but the great unknown was the devastating effect of wind forces.

It appears that until after the tragic collapse of the Tay bridge in 1879 few, if any, tests of wind pressures had been made. Designers were still content to adopt the figures presented by Smeaton to the Royal Society 120 years before (1759), which varied from 6 pounds per square foot for "high winds" to 12 pounds per square foot for a "storm or tempest". It is perhaps surprising to us—with hindsight—

that there appear to be no records of accurate wind pressure tests made in Brunel's day, because there had been ample evidence for years of the collapse of bridges and other structures as a result of the aerostatic and the aerodynamic effects of high winds.

By an aerostatic failure, I mean one—such as that of the Tay bridge—where the structure remains static and undisturbed until it is blown over bodily; aerodynamic failures, which were much more common, are those—such as that of the Tacoma Narrows bridge in 1940—in which the wind sets up increasing oscillations in the cables and deck of the bridge which build up until the deck is torn away.

Brunel's experiments and tests on models had shown him that the most efficient form of strut was that with a circular or oval cross-section. He had used struts of 9 foot diameter as the upper tubes, tied by the chains below with bracing inclined from the vertical, to carry each track of the 300 foot spans at Chepstow, but on the single track Saltash bridge he decided to use a wrought iron tube of elliptical cross-section, 16 feet 9 inches broad and 12 feet 3 inches high. Two advantages were gained here in that the greater width increased the lateral stiffness of the tube and also enabled the suspension chains and bracing on either side of it to lie in a vertical plane and still leave sufficient clear width for the railway track. Moreover, an elliptical section not only offers less obstruction to wind forces than a rectangular section, in that it tends towards a streamlined shape, but it is also strong in its powers of resistance.

Brunel did not see fit to provide any system of lateral wind bracing in the deck, no doubt because after careful consideration he placed all his reliance on the lateral strength of the oval tube. His faith was justified and, although the Saltash bridge is in a high and exposed position, it has successfully weathered all storms and tempests throughout its life.

The tubes were built up of plates generally $\frac{1}{2}$ inch or $\frac{5}{8}$ inch thick, each measuring about 10 feet by 2 feet, connected by means of rivets $\frac{3}{4}$–$\frac{7}{8}$ inch in diameter. At longitudinal joints the plates are lapped but at transverse connections double-covered butt joints are used. Running the full length inside each tube are six longitudinal stiffeners, consisting of plates 12 inches by $\frac{1}{2}$ inch, on edge, three at the top and three at the bottom (Fig. 40). Closed diaphragms are provided at the ends of each tube over the bearings. In addition there are annular stiffeners of plates 15 inches deep with edge angles, spaced at about 20 foot centres throughout the length of the tubes; these are additionally stiffened by two single vertical angles, and one double vertical angle on the centre line, for the full depth. The tubes are also stiffened laterally by means of 2 inch diameter solid iron cross-ties which increase in number towards the ends.

An interesting innovation of Brunel's was the drilling of a number of 2 inch diameter holes scattered over the under side of each tube and left permanently open, presumably to provide ventilation. This is most important in enclosed spaces in steel or ironwork as otherwise the condensation of moisture inside may present a serious problem.

Each cable or chain consists of two tiers of wrought iron links, the centre of the upper tier being 16 inches above that of the lower one. Each tier is built up of 14 links 1 inch thick or 15 links $\frac{15}{16}$ inch thick, side by side, in alternate panels, the links being interlaced and connected through the eyes by means of wrought iron pins 4 inches in diameter, screwed at the ends and secured by a nut. The links are 7 inches deep throughout their length except at the end connections where they are enlarged around the pins. The length of each link is 20 feet between centres of eyeholes with a tolerance of $\pm\frac{1}{50}$ inch. The pin-holes in the links were made $4\frac{1}{8}$ inches in diameter, so as to allow $\frac{1}{8}$ inch clearance for the pins. This would not be considered good practice today, but was no doubt done to avoid immense difficulties that would have arisen in erection if a close fit had been attempted.

The connections of the chains on to the ends of the tubes over the shore and centre portals presented a major problem, which was most ingeniously solved. The distance apart of the two sets of suspension chains on each side of the bridge is 16 feet 9 inches, which is exactly the same as the overall width of the tube. The sides of the oval tubes were therefore built up with additional thick iron web plates riveted parallel to them so as to present three intermediate walls and two outside ones taking the reaction from each pair of chains. The chain links were gathered into groups to pass into the four spaces between these reinforced walls for attachment, and the actual connections were made by means of wrought iron bolts 7 inches in diameter with a grip length of 2 feet (Fig. 40). The bolts connecting the upper and lower tiers of cables were staggered and located 2 feet or so from the ends of the tubes. Obviously this method of connecting the links presented severe maintenance difficulties as the spaces between the walls and the chains were inaccessible for efficient cleaning and painting. In recent years, however, the problem has been overcome by filling the spaces with non-oxidising paint which gives protection to the ironwork and remains plastic.

At the junction of the two tubes over the centre pier the tubes are

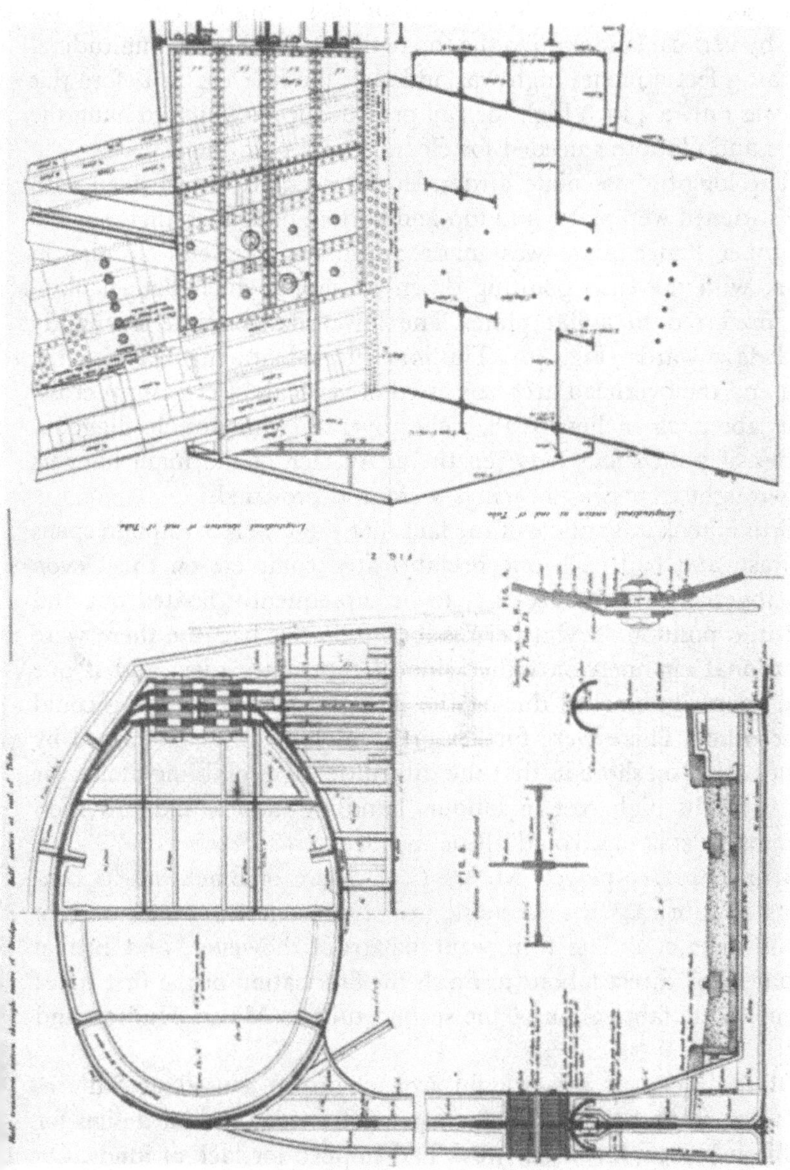

British Rail

Fig. 40. Royal Albert Bridge: junction of tubes and links at centre pier

massively strengthened by means of a complex arrangement of diaphragms and tie-bars. Easy access is provided into the tubes at the outer ends by means of ladders and manholes so that any authorised person can walk inside throughout their length. There is also access by vertical ladder on to the top of the tubes where a longitudinal handrail 3 feet 3 inches high was added in the late 1920s. Before this there was only a 3 inch high toe rail provided from which to hang the ladders and platforms needed for cleaning and painting.

The longitudinal plate girders in the deck are 8 feet deep with single stiffened web plates and top and bottom flanges 30 inches wide. The upper flange plate was made approximately semi-circular in section, with the ends pointing down—a shape which Brunel obviously preferred to a flat plate. The lower flange plate is slightly dished downwards (Fig. 40). The longitudinal bracing between the cables and the overhead arch consisted of verticals of cruciform cross-section about 24 inches by 18 inches overall, with double diagonal bracings of 7 inch flats between them. At each of the main hangers light wrought iron sway bracings were also provided.

Brunel took advantage of the fact that there were two main spans at Saltash and had each one prefabricated complete on the Devon shore, together with its decking, to be subsequently floated out and raised into position. It was permissible to do this because there were navigational channels on either side of the centre pier, and if one was temporarily blocked during the lifting of a span shipping could use the other. There were further great advantages to be gained by prefabrication on shore in that the difficulties of assembling aloft over water, with its high cost in labour, handling, access and provision of temporary stagings, could all be avoided.

Brunel first employed Messrs C. J. Mare of Blackwall as contractors to fabricate the wrought iron tubes on site, at a contract price of £162,000. The firm went bankrupt, however, and Brunel had to employ direct labour to finish the fabrication of the first tube; he then let the fabrication of the second tube to Messrs Hudson and Male.

Of the 2,800 or so wrought iron chain links used at Saltash, 1,150 were made before 1843 for use on Brunel's original design for the Clifton bridge, on which work had stopped for lack of funds. On all these links, which had been made at the Copperhouse Foundry, Hayle, at a cost of £19–10*s*–0*d* per ton, the eye ends had been welded to the 7 inch wide bars. The balance required at Saltash, consisting of some 1,630 links, were made in 1858 by Messrs Howard Ravenhill and Company, of the King and Queen Ironworks at Rotherhithe, and were each rolled in one piece (Howard, 1849). Brunel considered that the method of rolling the links in one piece was, no doubt, a great improvement, but he thought that with moderate care the welded links were equally strong. Most if not all of the rolled links had clamping cheeks at one end to provide a grip for handling, and every link was inspected and tested by Brunel's men during manufacture.

As soon as the fabrication of the first span was completed the span was freely supported at the ends and test loaded with 1,190 tons uniformly distributed throughout its length. The deflection at the centre of the span, due to this load in addition to the weight of the truss, amounted to 5 inches and the maximum stress in the wrought iron tube and chains was about 10 tons per square inch. As these results were considered very satisfactory, preparations went ahead at once for floating out the span. A temporary modification was made to the ends of the deck by raising the two end panels of plate girders 8 feet or so and moving them along a few feet towards the middle of the span. The purpose of this was to enable the pontoons to be brought in at a higher relative level and so reduce the height above water of the span when it was afloat.

On the day fixed for floating out—1st September, 1857—Brunel took personal charge from a vantage point high up on the span, his orders being conveyed by an elaborate system of flag signalling. Direct labour only was employed for the whole of the floating out and lifting of the spans and no contractors were engaged. Captain Claxton had been given command of five naval vessels, all ready at their moorings in the river. Brereton and Captain Harrison, the Commander of the *Great Eastern*, were with Brunel, and Robert Stephenson was only prevented from attending by a severe illness. It was a great day for Saltash—church bells pealed, the village was bedecked with flags, a public holiday was declared and crowds of onlookers flocked to the water's edge. Soon after midday, as the tide rose and the water was pumped out of the pontoons, a murmur of excitement arose from the crowd as the great span lifted slightly and was seen to float.

Brunel had planned and rehearsed the whole operation to be carried out within two or three hours at high tide, using the rising water to lift the span from the shore and the falling tide to lower it in place on the piers. It all went without a hitch. As soon as the span had risen 3 inches above its supports on shore, the pontoons carrying it hauled on their warps and brought it floating out to midstream. The span was swung majestically round to the Cornwall shore and warped by means of tackle into its final position. By three o'clock in the afternoon the pontoons were lowered in the water and as the tide fell they moved slowly out, leaving the great span supported at each end securely on the piers.

Then began the slow process of jacking the truss up from its then level—a few feet clear of the water—into its final position 98 feet up on the piers. The lifting was done by means of three hydraulic jacks located under each end of the span, the middle 20 inch jack by itself, or the outer two 11 inch side jacks together, being capable of supporting the whole reaction. The span was raised in 3 foot lifts, one end at a time, and after each lift the ends of the span had to be temporarily supported before the jacks were replaced and the next lift was made.

At the shore end of each span the masonry piers were built up beneath the ends of the tubes after each lift, but at the centre pier the two near octagonal columns and the cross-bracing between them were built up beneath the ends of the tubes to provide temporary support. To prevent the risk of accidents during jacking, the rams of the jacks had screw threads cut on them and the movement out of the rams was followed by a large nut screwed hard up. As an additional safeguard, timber packings were inserted, clear of the jacks, between the under side of the span and the top of the pier.

Ten months later the span had been raised to its full height and permanently fixed in position. Brunel had only been able to attend one of the lifting operations as he was busy supervising the launch of the *Great Eastern*; the work was left in the reliable hands of his deputy Brereton, who also took charge of the floating out and lifting of the second span as Brunel had to stay abroad on account of ill-health.

The second launch was accomplished as smoothly as the first (Fig. 41). Throughout the period of thirteen months occupied by the raising of the two spans, Brereton supervised the majority of the 140 lifts of 3 feet that had to be made. Speaking at a meeting at the Institution of Civil Engineers (Brereton, 1881) some twenty years later he said:

> "On no occasion was any interruption felt from the result of lateral wind-pressure, although winds frequently blew hard; and on one occasion, when the tube had nearly reached its full height, the wind blew a very strong gale, men were scarcely able to stand, and hats could not be kept on, the tube at one end resting on packings on the pier, and the other end poised upon the press rams which were run out to the full 3-feet stroke, not connected together, and with only 9

inches length of turned surface at the collars. No indications of excessive strain or nip were shown upon the guide slides which had been provided for such a contingency."

Brereton went on to criticise the then recent Board of Trade requirement that for a velocity of 110 miles per hour the design wind pressure would be 56 pounds per square foot, which of course has since proved to be nearly double the correct figure. His criticism was supported by no less an authority than Sir Benjamin Baker, who subsequently carried out the first accurate series of experiments to measure the force of wind of various velocities on large vertical areas, for use in the design and erection of the Forth Railway Bridge in 1881–90.

A further test was made on the Saltash bridge by Colonel Yolland, the Chief Inspecting Officer of Railways, a month before it was opened. A uniformly distributed load of 2.75 tons per foot was applied over the whole length of the bridge, under which the deflection at mid-span measured $7\frac{3}{4}$ inches on the western and $7\frac{1}{2}$ inches on the eastern span. Colonel Yolland estimated that under train loads existing at that time, which were normally not more than one ton per foot, the stress in the wrought iron tubes and cables would not exceed 4.2 tons per square inch. He considered these results "highly

British Rail

Fig. 41. Royal Albert Bridge: raising of second span

satisfactory and . . . greatly superior to anything of the kind which has been obtained elsewhere" (Berridge, 1969).

On 3rd May, 1859, amid scenes of rejoicing the bridge was officially opened by Prince Albert. A notable absentee, however, was the Engineer who had been the mastermind and the driving force behind the whole conception of the bridge. Worn out by the incessant strain and worry of all the major works he had accomplished, Brunel was very near the end. But he had a compelling urge to see the greatest of his bridges before he died. And this he did, lying on a couch on a specially prepared wagon of the Cornwall Railway, which was drawn slowly over the spans of his last great masterpiece, the Royal Albert Bridge.

The total cost of the bridge and its foundations, which had occupied some six years since work first started at site, amounted to £225,000—a very reasonable sum when it is compared with the £601,865 cost of Robert Stephenson's Britannia bridge. In fairness to Stephenson, his bridge carried two standard narrow gauge railway tracks of 4 feet 8½ inches, which would together be heavier than the single broad gauge 7 foot track at Saltash, but the overall length of the Britannia bridge was also 600 feet less than its rival and the presence of the Britannia Rock in midstream obviated the necessity for a deep water pier. Another reason for the economy at Saltash was, no doubt, that the much greater midspan depth of 56 feet reduced the weight of each span to little more than two thirds of that of each span at Britannia.

Brunel's fee for the work was a fixed charge of £5,000 which he was prepared to take in paid-up shares in the Company, together with a salary of £1,000 a year plus £200 a year for travelling. This was, of course, in addition to his office expenses, including the cost of surveyors, draftsmen and other staff engaged on the contract.

Figure 42 illustrates the view of the bridge at deck level. The appearance of the bridge and its approaches is improved by the fact that it is virtually free of any attempt at adornment for aesthetic reasons. The Board of the Cornwall Railway, after Brunel's death on 15th September, 1859, wrote in their report to the shareholders:

> "Considering the extraordinary difficulties which were overcome and the magnitude of the operation, it is believed that there is no engineering work in existence which has been more economically completed."

Today it is generally accepted that the four essentials in the design and construction of a bridge are safety, economy, durability and appearance—in that order—and in all these four qualities the

Fig. 42. Royal Albert Bridge at deck level

Saltash bridge cannot be faulted. For appearance, the bridge relies essentially on its simple and functional design, which creates an impression of strength and permanence. We must remember that tastes change and re-change through the years, and what is admired in one period may be questioned in another. As regards durability, it is expected today that the life of a bridge shall not be less than 120 years; by 1980 the Saltash bridge will have survived this span and let us hope will still have years of useful life ahead. But, as P. S. A. Berridge, an authority on the Chepstow and Saltash bridges, points out (1969), Brunel was too good an engineer to make a bridge much stronger than was necessary for the design loading. As the years go by the live load increases and the effects of fatigue and corrosion build up. This must inevitably tend to erode the factor of safety, especially in places that are most vulnerable, such as the pin connections of the links, where there are high tension and bending stresses and where there is little room for cleaning and painting.

Let us look now at the maintenance of the bridge and the details where modifications or strengthening have been carried out. It will be seen that these are confined to the deck, hangers and bracing, and are in most cases necessitated by the deflections and distortions consequent on the passage of heavy trains in both directions. In spite of the constantly increasing demands on them, however, the main tubes and chains remain virtually unchanged and have not been strengthened since they were built.

It appears that the wrought iron tubes and links were originally protected with a first coat of red oxide paint, followed by coats of medium grey. Subsequent coats consisted of red lead and white lead primer. On the outside of the tubes micaceous iron ore paint is now used and on the inside there is white paint on the upper half and black bitumastic below. The plate girders in the deck still have their original paint plus the layers added since. Cleaning and repainting have been carried out at intervals of not more than five years and grit blasting is used for cleaning where necessary.

The principal modifications to the trusses in chronological order are as follows.

(a) The 7 foot broad gauge was replaced by the 4 foot 8½ inch standard gauge in May 1892, which necessitated altering the timber decking from the longitudinal timbers of the broad gauge to the cross-sleepers of the narrow gauge, and strengthening it by doubling up the number of cross-girders.

(b) All the approach spans were renewed in 1928–29.

(c) Horizontal lateral bracing was added half way up the verticals throughout the span and much stronger sway bracing was incorporated between the hangers above the railway track by doubling up the existing bracing.

(d) In 1930 and again in 1960 a number of the diagonal members required tightening; the lower ends of nineteen were replaced and retensioned and modifications were also made to the four centre verticals. "Staybrite" bars were substituted for midspan hangers that broke in 1932 and the present articulated links were fitted in 1960.

(e) In 1963 severe corrosion necessitated the replacement of four intermediate hanger plates by means of auxiliary yoke hangers and the expansion bearings were renewed at the ends of the track girders.

Two details in the original design of bracing and hangers may fairly be criticised. In the pin connections at the ends of all the diagonals, instead of fitting tightly, the pins were intentionally made ⅛ inch less in diameter than the pin-holes. This inevitably weakened their bearing, reduced the efficiency of the system, and led to the pins working loose under the passage of the locomotives and requiring replacement.

The other faulty detail was that the hanger plates below the middle verticals were inaccessible for cleaning and painting. Rocked by movement of the catenaries under passage of the trains, they failed through the combined effects of corrosion and fatigue. Numerous breakages occurred but were temporarily made good to keep the bridge in service, until in 1960 the detail was completely changed. New pin-jointed hangers accessible for maintenance were fitted and other adjacent items replaced as necessary. A hallmark of good design is to avoid any places where water may collect and lie on the metal and also any spaces between metal surfaces too narrow to allow proper cleaning and painting.

In 1966 the British Railways Board instigated a careful enquiry into the ability of the two main spans of the Saltash bridge to carry much heavier four-axle vehicles with a gross weight of 100 tons and a length of 50 feet which were being introduced into service by British Rail. This reappraisal was made by the Engineering Research Department, at Derby, which had done a lot of work to assess the effect of fatigue on the strength of wrought iron.

A series of tests was made on a single span two-dimensional Perspex model 10 feet long to a scale of 1 to 50, and the results were compared with those of another series of full-scale tests made on the bridge itself. Multiple exposure photography was used to determine modes of distortion and a voltage pulse technique was used with the electrical resistance strain gauges installed on various members of the bridge. There was good agreement on deflections and strains in members below the chains, some of which were seen to be stressed above the fatigue limit for wrought iron of that age. In particular it was found that the verticals between chains and track girders were being distorted into an S shape by the heavy moving load. It was finally decided that the best solution was to fit 48 new diagonal members of steel to restrain the shearing action between the track girders and the chains. These diagonals were flat plates 7 inches by $\frac{5}{8}$ inch in section, designed with suitable connections at the ends so that they could be post-stressed after erection and at any time subsequently if necessary (Leeming and Whitbread, 1974). After their assembly on the bridge in 1970–71, tests showed that there should now be no risk of early fatigue failure under the proposed heavier loading. It must be remembered that it is reversals of stress more than variations that cause fatigue and in the two dominant parts of the bridge—the arched tubes and the chains—there will never be reversals of stress. The bracings and hangers are the members most susceptible to such risks and if the new diagonal members show the need for tightening at any time in the future, their design permits this to be done.

It is a tribute to the devotion and dedication of all the engineers concerned that Brunel's great achievement from the railway age should be so cherished and preserved today.

IX

Theoretical Work

PROFESSOR T. M. CHARLTON

There is substantial evidence that Brunel was abreast of the scientific principles of his times and their application to practical problems within a wide range of engineering activity. Although that aspect of his activities is perhaps determined most readily in relation to his structural engineering, his documents and contributions to meetings of learned societies give an impression of his wider scientific interests covering engines (steam and hot air), hydraulics and guns (including wire-wound guns) for military purposes. In all of these matters it is likely that he benefited from his French parentage and education, for it is undoubtedly true to say that France dominated applied science generally for several centuries prior to Brunel's birth. Latterly (in 1824) Sadi Carnot, under the stimulus of Watt's invention, had expounded reversibility, within the caloric concept (Cook, 1948), as the criterion of maximum efficiency of a heat engine (tacitly assuming the second law of thermodynamics later to be justified by Clausius), and Navier (1826) had published an excellent and highly original comprehensive treatise on the theory of structures containing the theory of bending (and the "middle third" criterion) and a correct quantitative treatment of statical indeterminacy, including continuous beams, for the first time (Todhunter and Pearson, 1886, 1893). Then, in addition to Coulomb's multifarious contributions to applied science, the French school of hydraulicians of the seventeenth and eighteenth centuries included Mariotte (also noted for research into strength of materials), Belidor, Pitot, Chezy, Borda, Bassut and Venturi, with the analytical support of D'Alembert, Lagrange, Laplace and Coriolis (Rouse and Ince, 1957). Moreover, the science of mechanics was approaching a state of development comparable with present day knowledge derived from Newton's laws, having benefited by the clarification and expositions of the seventeenth century French Academician Varignon.

Because Brunel is, it is believed, remembered first and foremost for his construction of the Great Western Railway system—a task which exhibited his outstanding excellence as a bridge and structural engineer—his theoretical work relating to it shall have pride of place here. In order to form a judgement of that work it is necessary first to consider in some detail the state of the art and science of bridge construction in Britain (and the world) during the first half of the nineteenth century.

At the time of Brunel's birth (1806) metal construction was in its infancy. The masonry arch was still the usual form for large bridges as indeed it had been for centuries. But the first major metal (cast iron) bridge had been built some 30 years earlier by Abraham Darby at Coalbrookdale in Shropshire and the first (small) metal chain suspension bridge had been built in 1741 at Middleton-on-Tees. Then a sizable metal chain suspension bridge was built in 1796 by Jacob Finlay at Jacob's Creek in the USA (Pugsley, 1968). These developments reflected the emergence of cast and wrought iron as a commercially and technically viable structural material (Straub, 1952). The structural principle, that of the arch and inverted arch, was essentially the same in these novel major works, and it was this principle which Brunel adopted for his major structural works. It is, though, noteworthy that cast iron beams of I section were developed during the latter part of the eighteenth century for use in beam-and-column construction for large factory buildings as well as for minor bridges (Pole in Jeaffreson, 1864). The relative weakness of cast iron in tension was recognised and allowance made by making the tension flange broader and thicker than the compression flange and/or increasing the depth of a beam (for use on simple supports) towards mid-span. A further development was the trussing of cast iron beams to relieve tension in the main material. That device simply used wrought iron rods fastened to the ends of the beams and strutted out underneath to give the appearance of a partially triangulated truss. The trussed beam was used for some of the early small railway bridges (e.g. near Tottenham, by Stephenson and Bidder in 1839), and the disastrous failure of such a bridge by Robert Stephenson at Chester in May 1847, with the loss of five lives, resulted in the Royal Commission to Inquire into the Application of Iron to Railway Structures in August 1847. Among the members of that Commission was Willis, Jacksonian Professor of Natural Philosophy at Cambridge, who, along with G. G. Stokes, contributed a paper on the deflection of beams under moving loads for the benefit of the Commission. Brunel gave evidence to the Commissioners (Jeaffreson, 1864) whose report was published in 1849.

Although the first rolling mill for wrought iron flat bar and plate was commissioned in 1787 in Britain by Henry Cort (Straub, 1952), wrought iron bridge trusses of significance did not appear seemingly until the middle of the nineteenth century. Wrought iron plate was, however, used earlier, for the construction of flitched beams (with timber, by Smeaton), plate girders and tubes, the culmination of which may be said to be Robert Stephenson's Conway and Menai tubular bridges of 1848–50. It is likely that the relatively late development of the open bridge girder or truss was due to a combination of the problems of safety and economy. While the theory of the strength and bending of beams was established (finally by Navier) before the railway era, the theory of frameworks or trusses was not, in spite of the essential treatment published by Navier (1826). The graphical methods due to Rankine (1858), Maxwell (1864a), Bow (1873), Jenkin (1869) and Chalmers (1881) which brought the analysis of trusses within the competence of the engineering profession generally were not to emerge until about the time Brunel died. Also it seems that rolled wrought iron sections other than flat bar were not generally available and so the alternative of sections built up by bending and riveting would have been very expensive. According to Straub (1952) angle sections were first produced by rolling in 1830 following John Birkinshaw's patent for rolling rails for railways in 1820. The quantities of rolled sections were apparently small until the 1840s—when I sections became available, especially in France. A similar impression is given by Dempsey (1864).

Within the constraints imposed by the state of knowledge and technology of his times Brunel appears to have made the fullest use of scientific principles and experiment in the design of his structures. Evidence of that is to be found in his calculation books, sketch-books and letters, which contain detailed calculations relating to the design of arch and suspension bridges in addition to particulars of an experimental approach to the behaviour of continuous beams. Although he did not publish particulars of his work in the form of books or papers, Brunel was an avid contributor to discussion (recorded in print) at the Institution of Civil Engineers and there exists a published account of one of Brunel's continuous beam experiments in 1849 which is considered at length elsewhere in this chapter.

The method used by Brunel for the design of arches originated, it seems, in the seventeenth century and is variously ascribed (W. H. Barlow, 1846; Straub, 1952) to Hooke, De la Hire, Parent and David Gregory. Timoshenko (1953) asserts that the first application of statics to the solution of arch problems is due to De la Hire in whose book, *Traité de Mécanique* (1695), the funicular polygon is used for the first time to analyse an arch. Also Straub, referring to De la Hire's book, asserts that it was argued that the shape of an arch must be such that for each voussoir the resultant of the dead weight and the pressure of the preceding voussoir is normal to the face of the next voussoir (known by Brunel and his contemporaries as the "wedge theory"); friction was neglected. Moreover Straub seeks to identify Gregory's contribution with that of De la Hire, stating that the discovery is "summed-up by the Scottish Mathematician, D. Gregory in his treatise *'Properties of the Catenaria'* (1697), in the theorem that the theoretically correct central line of the arch must be shaped like an inverted catenary". Details of Brunel's arch calculations using the wedge theory are given in Chapter V and include approximate representation of the distribution of shear force in an arch by a polygonial function followed by integration to provide the relevant thrust line equation.

Brunel's faith in elementary statics was apparently undisturbed by the Reverend Professor H. Moseley's[40] attempt to improve arch analysis by what he enunciated in 1833 (Moseley, 1833, 1835, 1843) as the "Principle of Least Resistance (or Pressure)",[41] and W. H. Barlow's attempt (1846) to adapt it for use in engineering practice. The theoretical improvement sought took cognisance of friction between stones or voussoirs and it is noteworthy that Barlow referred to Coulomb's relevant work (1773) as well as results obtained by William Whewell (mathematician, author of books on practical mechanics and, subsequently, Master of Trinity College, Cambridge).

It is salutary to recall Brunel's contributions to the discussion of Barlow's paper (Brunel, 1846). First he remarked:

> ". . . the compressibility and elasticity of materials of construction had not been sufficiently insisted upon. This did not generally obtain enough consideration, yet it was of great importance to the stability of a structure; all materials, even granite, possessed an amount of elasticity, and it did not suffice to have the line of pressure fall merely within the mass (as Moseley's theory seemed to indicate); it should be sufficiently

within it to allow for any yielding from elasticity, without endangering the building."

Brunel's letter in his private letter book *(PLB, 30 Dec., 1854)* to W. Bell repeats this viewpoint. Secondly, consequential to remarks of G. P. Bidder that he:

"... was tempted to consider an arch constructed by rectangular bricks set in a matrix of cement, as a bent trussed girder, the tension rods of which were represented by the abutments of the arch. Very flat arches such as the Maidenhead Bridge, were examples of what he meant"

Brunel said that he:

"... could not agree with Mr. Bidder's comparison, or what he might be permitted to term his amusing theory; on the contrary, he must contend that there was no analogy between the arch and a trussed girder. In the former the main force was pressure, in the latter the force exerted was tension; the abutments of the one had to resist a horizontal thrust, at a given angle, whilst the wing walls, under the other, had to support only a vertical pressure; any tendency towards horizontal thrust, which might have arisen from deflection of the beam or girder, was prevented by the tension rods which connected the opposite extremities. If an arch could be considered as a bent trussed girder, it must follow, that it would stand equally well whether the curve was upwards or downwards, which certainly did not accord with his notions of the properties of an arch."

Thirdly, after much more discussion by the eminent engineers present and immediately following a contribution by Robert Stephenson to the effect that:

"... mathematicians always considered the line of pressure to be at right angles with the supporting surfaces or the abutments. It would appear, however, from Mr. Barlow's explanation, that instead of drawing a series of lines at right angles to the surfaces through given points, thus forming what might be termed the Polygonal theory, he described a correct curve through the same given points. Mr. Stephenson could not understand how the voussoir could give a line differing from the line of force treated of by mathematicians." (Stephenson, 1846).

Brunel said:

".... the subject was one of great difficulty, as it embraced so many considerations; it might, however, he thought, be rendered simple, by considering an arch not as composed of separate voussoirs bound together by cement, thus involving other principles, but as a homogeneous, and, he might almost say, an elastic mass. If viewed in that light, the pressure would be found to extend more or less over the whole surface. The 'centre line' or 'neutral axis' might in such a case receive the denomination of the 'line of pressure'. If this idea were followed-up, there would be less difficulty in explaining the principles laid down by Mr. Barlow."

The subject of stability of arches was considered also in a paper to the Institution of Civil Engineers by G. Snell (1846). Again Moseley's theory was invoked and it was noted that Lamé and Clapeyron had provided a similar theory by different means. Brunel did not contribute to the relatively brief discussion but Robert Stephenson did so and it is very worthwhile recalling some of his remarks, which indicated that neither Snell nor Barlow had provided an acceptable alternative to the wedge theory for engineering practitioners. Thus Stephenson said:

"... it might be said of this, as of Mr. Barlow's paper, that it was not so practically useful as it might have been, had it treated of the stability, instead of the instability of, of certain arches. The problems given, hardly met the wants of the engineer. It appeared, that examples of arches nearly in a state of equilibrium were selected, instead of investigating the more abstruse question, of an arch unequally balanced. The old theories all viewed the arch in the same manner, and had for object, to establish when the acting force had not a tendency to move. These old theories and formulae were looked upon by Mr. Stephenson with great respect, and acting upon them, he had been generally successful. The investigations of Professor Moseley, Mr. W. H. Barlow, and Mr. Snell, were most important, but Mr. Stephenson must own, that he had studied attentively the elegant and scientific investigations of the former of these gentlemen, without clearly understanding their practical application" (Stephenson, 1846).

It is interesting that Moseley did not apparently attend and contribute to the discussion of the papers by Barlow and Snell respectively. Moseley's approach embodied the assertion that the line of pressure of an arch was determined by the condition of minimum horizontal thrust at the crown. That had to be achieved in practice apparently by trial and error. Perhaps the evidence of the longevity of Brunel's (flat arch) Maidenhead bridge is adequate testimony to the practical viability of the old wedge theory for the purpose of designing voussoir arches. The general problem of the arch was to be solved later in the nineteenth century using the theory of elastic behaviour (expounded qualitatively by Brunel) by Bresse in 1854[42] (according to Todhunter and Pearson (1886, 1893) and Timoshenko (1953).

Of Brunel's application of scientific principles to the design of suspension bridges, there appears to be no evidence that he did other than use the inversion of the wedge theory which was the basis of his

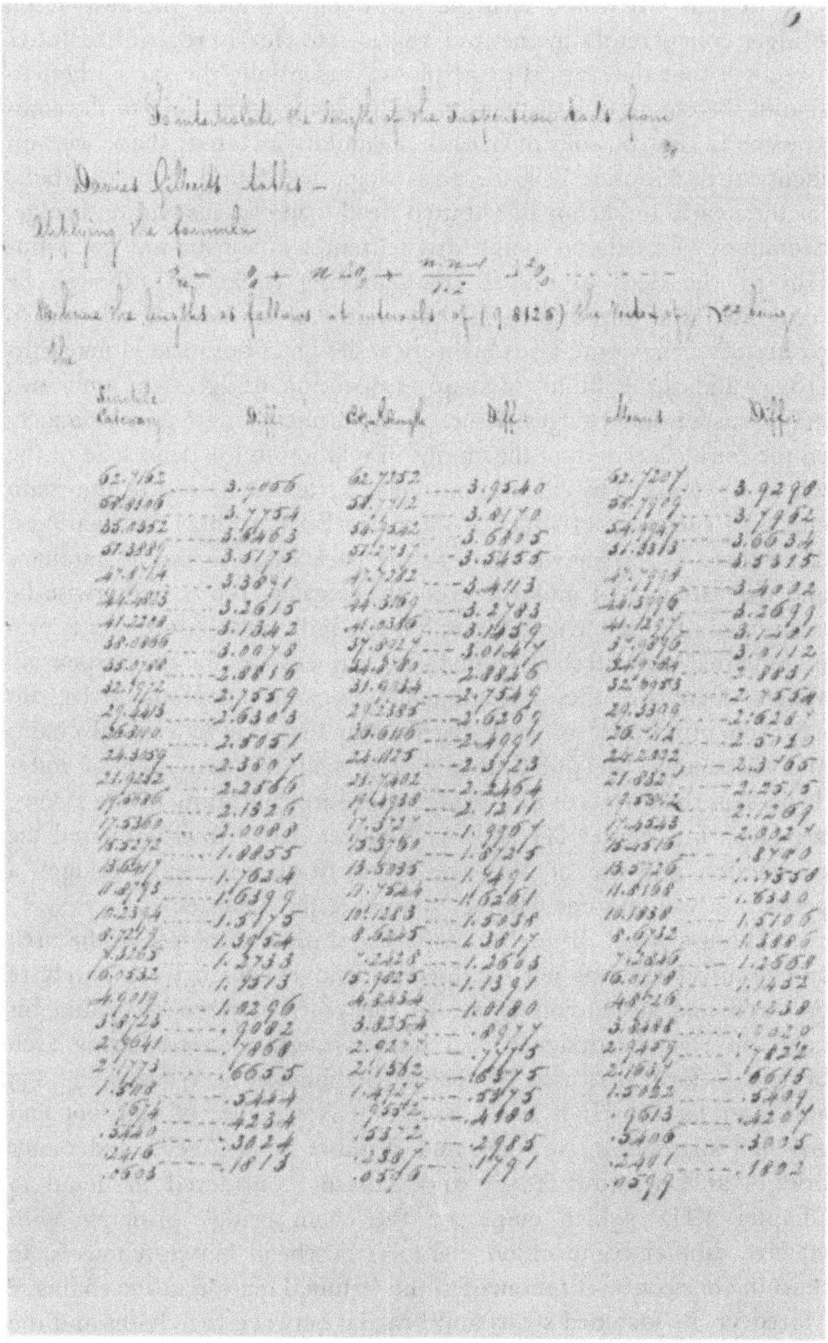

Fig. 43. Clifton bridge: specimen of calculation

arch designs. (It would seem he was designing arch and suspension bridges concurrently in the mid 1830s.) He clearly recognised (after Gregory) that the statical principle was essentially the same, the tension in the chains of a suspension bridge being equivalent to the compression in the voussoirs of an arch of similar (inverted) shape with an identical distribution of load. That shape was found to be parabolic for the nearly uniformly distributed dead load of a suspension bridge. Seemingly he made no quantitative attempt to ascribe any contribution by the deck to overall load-carrying capacity—although he recognised the importance of that source of stiffness in relation to oscillatory behaviour. Thus his statical design of both the Hungerford (1843) and the Clifton (1829–40) suspension bridges (the only two major suspension bridges designed by Brunel) was based primarily on the tensile strength of the chains in relation to the dead load of the whole structure. His calculations relating to the design of the main chains, a specimen of which is shown in Fig. 43, show an awareness of all three of the known shapes—the parabola, the simple catenary and the catenary of uniform strength (Pugsley, 1968). Otherwise he performed strength calculations for the individual components and investigated the effect of changes of temperature on the suspension system. Details of these bridges, including the ingenious devices for ensuring uniformity of load distribution between individual chains and suspenders for the Clifton bridge and the provision of roller devices on the towers of that bridge to ensure continuity of the chains, are given in Chapter III. It is well known that Brunel witnessed the completion and use of only the Hungerford suspension bridge—a pedestrian way of some 676 foot span over the Thames.

The use which Brunel made of the statical principle of the arch and its inversion was not confined to bridges which were purely of the arch and suspension types. In the years between 1845 and his death in 1859 he designed and built wrought iron bowstring arch bridges (a form first used by Robert Stephenson at Weedon in 1835 and later for the High Level Bridge at Newcastle) of 100 foot and 200 foot spans near Newport and Windsor respectively, and major bridges at Chepstow (1852) and Saltash (considered in detail in Chapter VIII) which employed the chain bridge principle with massive tubular compression members overhead between towers, to resist the horizontal component of the terminal tension in the chains.[43] Moreover, he included struts and bracing between the chains and the tubes to provide additional stiffness to the bridges, designed as they were for the relatively high velocity live loading caused by railway

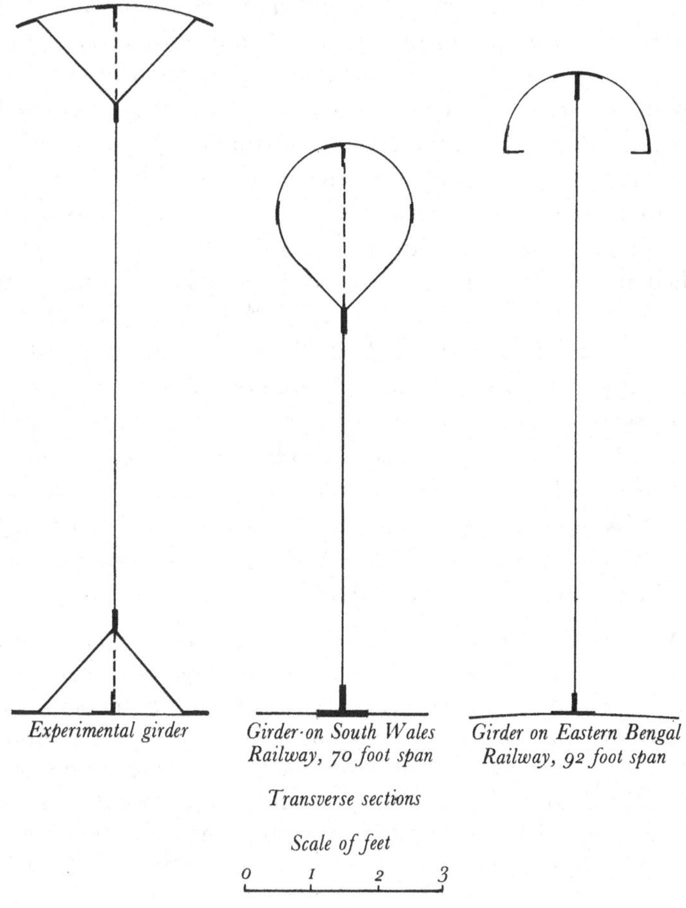

Fig. 44. Plate girder sections

traffic. (The Saltash bridge with its two identical main spans which incorporate essentially both the suspension chain and the arch was completed only three months before Brunel died.) It should also be recalled that in the early 1840s Brunel carried out large-scale experiments on the strength of plate girders and used novel forms of compression flange including the tube and variants thereon (Fig. 44).

The design of the Chepstow bridge stimulated a particularly noteworthy theoretical investigation by Brunel and his assistant W. Bell. It concerned the analysis of continuous beams, an essential ingredient of the Chepstow bridge as well as the contemporaneous large tubular railway bridge over the Menai Straits (for the design and erection of which Robert Stephenson was helped by Brunel as well as Fairbairn and Hodgkinson) and the somewhat different though sizable railway bridge with tubular girders spanning the River Trent at Torksey (Fairbairn, 1851). In addition to the main span with its suspension chain, the Chepstow bridge included four lesser spans and, in common with distinguished contemporaries, Brunel was desirous of obtaining the economic benefits which were known to accrue from continuity. That knowledge was seemingly largely due to Moseley's excellent book published in 1843 which described much of Navier's theory and analysis of beam systems as well as Poncelet's use of strain energy.[44, 45] Thus it contained a correct analysis of encastré and continuous beams using the compatibility approach wherein the equations of statics are supplemented by those relating to continuity of slope at intermediate supports, to enable a sufficiency of equations to be obtained for precise analysis.[46] Moreover, Moseley's treatment of the subject emphasised the benefits of continuity with regard to stress and deflection. In his paper of 1871, "On the Stresses of Rigid Arches, Continuous Beams, and Curved Structures", Bell relates that:

> "While examining, in the year 1849, the stresses on continuous beams for the late Mr. Brunel, in reference to the large bridge at Chepstow, the lower girder of which is virtually a continuous beam of five unequal spans, the Author, using the method of Navier, found the subject one of no inconsiderable difficulty, from the number and complexity of the eliminations required; and he was gratified by the formulae at which he then arrived, for beams up to five spans, being completely verified by the experimental tests to which he subjected them, and which were devised by Mr. Brunel."

It is noteworthy that the discussion of Bell's paper included contributions from Clerk Maxwell, Fleeming Jenkin and the Frenchmen Collignon and Gaudard. It was a fascinating discussion in which attention was drawn not only to the work of Bresse but also to that of Culmann and to the use of the principle of virtual velocities for determining the deflection of structures by Clapeyron and Lamé in 1852. Indeed it was Bell in his reply to the discussion who described the latter, having remarked upon Maxwell's use of Clapeyron's theorem—conservation of energy—in his celebrated paper on statically indeterminate frameworks (Maxwell, 1864b). It is, incidentally, not unreasonable to infer that Bell's distinction in engineering science was due in large measure to Brunel's influence in the formative stages of his career.

The genesis of the tests (Fig. 45) to which Bell referred is to be

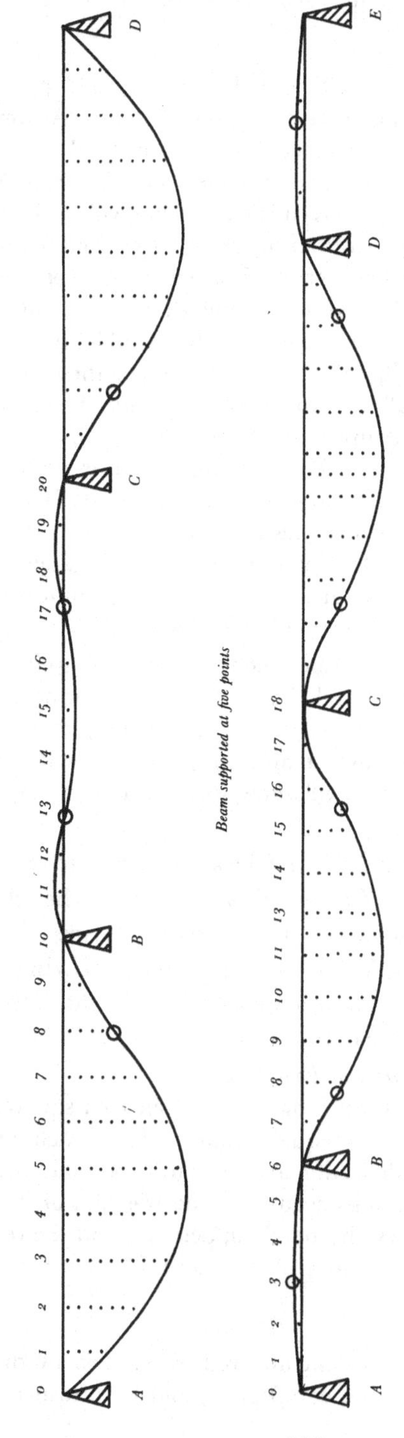

Fig. 45. Continuous beam experiment

found in Brunel's letter to Bell *(PLB, 12 May, 1849)* which reads as follows:

"The point to be investigated both theoretically and practically i.e. experimentally which most presses at present is this. Assume a beam supported only at 4 points and then discontinued—and not a continued beam as before assumed—assume it loaded and then assume the two points B & C to be lowered a given extent—what will be the law for the increase of strain. It may be rather dangerous to depend upon the result of an investigation into the General law and it may be for the same reason, dangerous to depend upon any experiments in which the dimensions and circumstances do not closely resemble the case to which I want to apply the result, and therefore I give you these circumstances and dimensions and shall be quite content to have the results for that case only, and not to attempt to generalize too much.

"The case is this, a beam 7 feet deep supported as under is loaded continuously (but in such proportion that the points B & C remain horizontal) to an extent that strains the upper and lower web to 5 tons per inch or an extension of 1/2000, what will be the alteration in the nature of the strains, and what the increase at any point—if the point of support at B & C are lowered 1″ and what at 2″.

"It will perhaps more clearly resemble the state of things if we say that the beam is thus supported and that the load is uniform and that the three points B C D are lowered 1″ & 2″. You may at once try this (without of course seeking to resemble the 7 ft depth or the strain) but the character of the change will show itself at once if you gradually lower the three points."

Again, in the book by Edwin Clark (assistant to R. Stephenson), *Britannia and Conway Tubular Bridges* (1850), the chapter on "Deflection of Continuous Beams" (in which the analysis of continuous beams was contributed by W. Pole), gives the following details of the experiment (Fig. 45) communicated by Brunel and accompanied by theoretical formulae and calculations.

"*A Continuous Beam supported at Four Points*

"A rod of yellow pine 38 feet long, and half an inch square, was supported on four bearings at equal distances from each other, in a horizontal line, and the deflection and pressures upon the various points of support were carefully observed. The rod was laid on the supports with each of the four sides alternately uppermost, and the results given below are the means of the four observations. The curve is plotted in the accompanying Plate.

"*Deflection*

"The deflections of the experimental rod were taken by dividing the length of each span into ten equal parts, and by measuring, at each division, the distance of the under side of the rod from a straight-edge fixed below.

"In the following table column 1 contains the points of observation, column 2 the mean reduced deflection in inches as observed, and column 3 the calculated deflection.[46]

	Points of Observation (or value of x/l)	Observed Deflection	Theoretical Deflection (or value of y)
		Inches	Inches
Support on the outer prop	.0	+0.000	+0.00
	.2	+1·422	+1.38
	·4	+2·08	+2·06
Centre of outer span	.6	+1.77	+1.83
	.8	+0.87	+0.89
Support on the inner prop	1.0	+0.00	+0.00
	1.2	−0.09	−0.08
	1.4	+0.07	+0.12
Centre of middle span	1.6	+0.07	+0.12
	1.8	−0.09	−0.08
Support	2.0	+0.00	+0.00

"*Pressures on the Supports*

"The pressures on the supports of the model beam were measured by a steel-yard, applied at each bearing of the beam in succession. The distance from the fulcrum to the point of support of the beam was always 18 inches, and the weight on the steel-yard half a pound. The leverage, or distance of this weight from the fulcrum, was carefully recorded as the lever was applied to the different points of support.

"The mean weight on each of the outside supports A and D, in terms of the leverage as above, was 7.53; the mean weight on each of the middle supports B and C was 21.78; giving the total weight of the rod

$$=\frac{1}{2}\times\frac{58.62}{18} = 1.63 \text{ lbs.}$$

"The absolute values, therefore, of the pressures given by the experiment are as follows:

	lbs
On support A ... =	0.210
,, ,, B ... =	0.605
,, ,, C ... =	0.605
,, ,, D ... =	0.210
Total weight of rod	1.630

"The calculated pressures may be obtained from . . . we have the following values:

		lbs
On support A	... =	.217
" " B	... =	.598
" " C	... =	.598
" " D	... =	.217
		1.630"

Although the lattice type of construction—i.e. a flanged girder with a complexity of crossed flat wrought iron bars for the web, representing a potentially physically lighter form than that with a solid plate web of the same thickness of metal—had been introduced during the early 1840s (e.g. a bridge on the Dublin and Drogheda Railway, some 3 miles from Dublin, described by G. W. Hemans before the Institution of Civil Engineers in the session 1843–44, during the discussion of which it was said that, according to Dr Gregory in his book *Mathematics for Practical Men*, the original inventor of the lattice bridge was Mr Smart), it did not seemingly find favour with the distinguished designers of major structures such as Stephenson and Brunel. Among its more reputable advocates, was W. T. Doyne who describes (1851) such a bridge for carrying a public road over the Rugby and Leamington Railway at the Honingham cutting. Referring to the calculation of its strength he seems to rely on the theory of bending of beams, asserting, by comparison with a conventional plate girder:

> ". . . the only difference being, that the connecting link between the top and bottom, whose duty it is to carry the strain from one to the other, instead of being one continuous rib, has several points of distinct attachment. And as the amount of strain thrown into the top and bottom throughout the whole length, or at any point, is in both cases the same, it is evident, that the whole action upon the connecting links is the same and that the strain upon each lattice may be calculated: this strain is not uniform throughout, but depends upon the position of the load, so that it is necessary to calculate the proportions of the lattices for each part, or to make them all of the maximum strength."

It is noteworthy that Doyne does not mention shear forces specifically, and therefore his explanation of the calculations relating to the proportions of the lattice work is less than satisfactory—a matter which drew adverse criticism during the discussion from C. H. Wild and G. P. Bidder. The latter further asserted that he believed the construction was less economical than that embodied in the conventional plate girder.

Subsequently Brunel (1851) entered into discussion of a paper by W. T. Doyne and Professor W. B. Blood (1851), of University College, Galway, concerning strains on the diagonals of lattice beams. On that occasion it was the Warren girder (named after its inventor, Captain Warren) to which they addressed themselves: a very different proposition from the lattice girder formerly described by Doyne, as Brunel noted by saying, "it was necessary to draw a distinct line of demarcation between the lattice bridge and that kind of construction called Warren's girder; in the former much of the material employed was useless, whilst in the latter, if properly proportioned, every part was made to perform its duty, either bearing pressure, or in tension; he was inclined to think it might be rendered nearly the most economical, as well as the most efficient kind of girder, for spans of a certain extent." It seems that Brunel was clearly mindful of the effects of self-weight which he and Stephenson had offset for long spans by the device of continuity and tubular construction, first used by the latter in 1847 (Pole in Jeaffreson, 1864). His assessment of the Warren girder was clearly accurate. It was at that time novel, being first used (according to Pole) in 1850 at the London Bridge station. Later, in 1852 it was used as developed by C. H. Wild (Cubitt, 1852) for a span of 240 feet to carry the Great Northern Railway over a tributary of the Trent near Newark while in 1857 it appeared in the Crumlin viaduct in Monmouthshire. It is, perhaps, worth recalling that Hawkshaw contributed to the discussion also, asserting that he had constructed many lattice bridges of various spans, including the Paddock viaduct near Huddersfield with continuous malleable iron latticework, designed (presumably in the manner outlined by Doyne and Blood) in 1850.

In fairness to Doyne and Blood, however, it must be recorded that they presented descriptively an accurate approach to the calculation of the forces in the members of their lattice (Warren) girder; namely that of analytical resolution of forces at joints. That approach was clearly appreciated by Brunel so it is of particular interest that even before the advent of graphical methods leading engineers had used elementary statics to assess correctly the forces in some relatively simple kinds of bar structure. It is likely that Doyne and Blood gave the first account of that to be published in Britain and, moreover, justified their findings by "dynamometer" measurements on a small-scale model. In their enthusiasm, though, they committed the error of asserting that the same principles could be used for determining the forces in the members of any latticework and it seems to have been that assertion which Brunel rightly challenged in a gentle manner. It should be recalled that Brunel himself used the bar type of construction in his timber railway viaducts, for the design of which it seems likely that he made use of elementary analytical statics having regard to Navier's extensive treatment of such structures (1826). Indeed, in 1846 (I. Brunel, 1870) he carried out large-scale tests on struts of yellow pine to investigate the reliability of strength calculations, including those by Euler's theory.

It might, perhaps, be said that the advent of metal "open girder" or truss construction coincided with the completion of the major wrought iron tubular bridges conceived during the 1840s. That impending development probably influenced Brunel in his choice of form for the Chepstow bridge and later for that at Saltash, being in a sense a compromise between the suspension, truss and tubular varieties. No doubt mindful of the costly preliminary tests which had been necessary for the design of major tubular bridges, as well as the apparent weight-saving inherent in the open girder, amenable to statical analysis of strength, that latter form was seized upon eagerly by Brunel's and Stephenson's successors (notably Thomas Bouch assisted by R. H. Bow) using the rolled sections available then. But fashions are transistory and the tube was to become established as a safe and economical form for major structures as well as large structural members.

An aspect of Brunel's theorising relating to structural strength in the broadest sense is his proposal of 1854 (I. Brunel, 1870) for strengthening the barrels of cannons by external winding with wire under tension (i.e. prestressing). He was disappointed when, in 1855, that idea was patented by one, Longridge, whom Brunel suspected of plagiarism; he was further irritated because he believed that the patent law was contrary to the interests of progress. (It is noteworthy that another idea of Brunel's concerning guns, namely the use of an octagonal bore with "spiralling" for small arms, had been anticipated by Whitworth.)

A discourse about Brunel's theoretical work would be incomplete without reference to his knowledge of the theory of prime movers and propulsion as well as the mechanics and behaviour of fluids and friction relating to the launching of ships.

Concerning prime movers, Brunel was for a period an ardent advocate of the atmospheric system of traction for railways. That was apparently founded on the belief in a general principle that stationary power, if freed from the weight and friction of any medium of communication, such as a rope, must be cheaper than locomotive power. He considered the waste of power which arises from the locomotive having to move itself as well as the train, and the excess cost of power supplied by a locomotive over that supplied by a stationary engine. Data relating to the former were available from experiments conducted by Daniel Gooch during the gauge controversy during the mid 1840s. His calculations (I. Brunel, 1870) involved:

"Total tractive force in lbs (F)
Weight of engine and tender (superfluous load) in tons (50)
Weight of train (useful load) in tons (W)
Resistance of machinery, rolling resistance and air resistance of engine and tender in lbs (R)
Rolling resistance of train in lbs per ton (K)
Gradient (G)
Speed in miles per hour (V)
Frontage area of train in square feet (63)
Frontage area of portion of train unprotected by the tender, in square feet (24)
Resistance of air in lbs (according to empirical formulae), (frontage area) $\times V^2/400$"

The calculations were carried out thus:

"For a locomotive train, $F = R + WK + 24V^2/400 + (50+W) + 2240\ G$;
For a train without locomotive, tractive force $= WK + 63V^2/400 + 2240\ WG$;
Useful tractive force $= W(K+2240G) + 0.1575V^2$
Tractive force wasted by using a locomotive $= R + 112000G - 0.0975V^2$
Hence: $F = (R + 112000G - 0.0975V^2) + W(K+2240G) + 0.1575V^2$
$W = (F - R - 112000G - 0.06V^2)/(K+2240G)$"

It was deduced that on a level track, at 40 miles per hour one third of the power usefully employed on the train was wasted on the locomotive, while that fraction became greater than one half at 60 miles per hour. (Gross horsepowers of up to 784 were quoted.)

Although Brunel's theoretical appreciation of the use of stationary power embodied in the atmospheric system was apparently substantially accurate, the system failed owing to mechanical defects involving substantial air leakages. (A detailed description of the atmospheric system is included in Chapter IV.)

In 1847 Brunel contributed to the discussion of W. Froude's paper on discharge of elastic fluids under pressure, which had a bearing on the atmospheric system of propulsion. Another contributor to that discussion was John Scott Russell who, with Froude, had been associated with Brunel in his design and construction of ships including the *Great Eastern*. Russell was a shipbuilder interested in theories of resistance of ships with particular reference to the influence of wave formation, and introduced a reverse curve form of bow as early as 1834. He was a founder of the Institution of Naval Architects. William Froude conducted tests on the rolling and resistance of the *Great Eastern*. Reference may be made to Chapter VII for an account of their remarkable activities in the design and construction of ships. But it is appropriate here to mention the careful experiments which Brunel conducted into the friction of ships in relation to launching, full details of which are to be found in I. Brunel's biography (1870).

Brunel also contributed to the discussions of Ericsson's heated air engine (Cheverton, 1853) and Poingdestre's paper (1849) describing Sir George Cayley's hot-air engine. The former relates to a regenerator device proposed by the Swedish engineer John Ericsson (inventor of the vane propeller for ships patented in 1836):

> "Mr Brunel, V.P., agreed in considering the regenerator to be a mystification, and the difficulty of the matter arose from its plausibility. It was extremely difficult to disprove that which did not exist at all. He believed the stated gain of power, from the action of the regenerator, to be mere assumption, and he was inclined to regard it just as he would any attempt to produce perpetual motion; still he admitted the difficulty of exposing the fallacy, as he contended it to be, when it was asserted, that the power derived from the expansion of air by heat, could be used effectively, and then be recovered and used again.
>
> "He could not gather, from any of the statements, how power was actually obtained; and he could only arrive at the conviction, that if there was any development of power, in lending heat to the metallic webs and borrowing it again, the natural consequence must be, that after a time there might be such an accumulation of borrowed heat, as would enable the machine to work, without any fire under the heating vessels. It was scarcely worth while to expend time in the refutation of such a fallacy. The same kind of error had been fallen into, with respect to the hot water from the condenser of a steam engine; but in that case it was clear, that the advantage arose, not from recovering heat which had been previously used, but by obtaining some that had not been used before, and which would otherwise have been wasted.
>
> "He was of the opinion, that Stirling's engine was not only a prior introduction, but that it was a better machine than Ericsson's; still he thought that neither of them could be advantageously compared with a steam-engine."

Referring to Cayley's engine Brunel said (1849c):

> ". . . he remembered some extensive experiments being conducted many years ago, by the late Sir Isambard Brunel, for the celebrated M. Montgolfier, who came to England about the year 1815, fully impressed with the feasibility of the scheme, which, however, although tried exactly in the manner recommended by Mr. Gordon, applying the expansion directly to the column of water, without the intervention of any machinery, proved a total failure. The vessels in which the fuel was consumed were of considerable height, and 8 feet or 10 feet in diameter, so that large volumes of heated air could be employed. The apparatus raised large volumes of water, 20 feet or 30 feet in height, but the result of all the experiments demonstrated, that the amount of caloric in gaseous bodies, was not greater than that produced by the expansion of water into steam, and that practically it was not so generally applicable. The researches of Sir Humphry Davy and Mr. Davies Gilbert confirmed this result.
>
> "The experiments on the application of the mechanical power of carbonic acid and other gases, condensed under pressures, which were extended at times to three hundred atmospheres, were continued for nearly fifteen years, at an expenditure of upwards of £15,000; although at first, and in theory, it appeared a beautiful means of obtaining power, the conclusion arrived at was, that commercially it was not so advantageous as that derived from the expansive forces of steam; and it must be evident, that the mechanical power obtainable from the expansion of common air, was much less advantageous than that of the more subtle gases. It appeared then, that although there was little doubt of the mere mechanical difficulties being overcome, there was reason to doubt the application of air engines being more successful at present, than at former periods."

Brunel observed further:

> ". . . that he quoted the experiments on condensed carbonic acid gas expressly as bearing on the question, and could assure Mr. Gordon, that he had succeeded almost without difficulty, in making joints perfectly tight, under pressures of 1,000 lbs and 1,500 lbs per square inch; the experiments he alluded to, for determining the amount of mechanical power, were not made through the intervention of machines where there would be friction and other causes of error, but a simple apparatus, free from these objections, in which the expansion of the gas was made to act upon mercury in another vessel, which was allowed to escape under a certain pressure, and the caloric absorbed by the expanding gas carefully measured, and the results deduced by recompressing the gas into a liquid and measuring the heat given out in the operation."

The foregoing quotations typify Brunel's broad appreciation of scientific principles. An outstanding aspect of Brunel's theoretical work is that it was undoubtedly stimulated entirely by his needs as an engineering practitioner embracing, in his day, almost every kind of engineering activity. In common with his distinguished contemporaries Stephenson and Fairbairn, he seemingly made extensive use of the experimental approach. He did not theorise for its own sake after the fashion of many academics and seemed disinclined to dissipate his energies in writing and presenting papers to learned societies. (Although a member of many societies, it seems he confined his attendance of meetings to those of the Institution of Civil Engineers.) Perhaps it was consciousness of this attitude of his father that caused his elder son, Isambard, to assert (1870) that:

> "Mr. Brunel was thoroughly conversant with the principles of mathematical analysis, and was able with great readiness to apply it in practice; but at the same time he preferred when it was possible, to use geometrical methods of solution for engineering problems."

While it cannot be said that Brunel contributed to science in principle, his mastery and application of science were clearly superb and a wonderful example to engineers of his own and future generations.

Notes

The engineering units used throughout this book are those current in Brunel's lifetime. They may be converted to SI units as follows.

Length	1 foot = 12 inches = 0.3048 metres
Force	1 ton = 2240 pounds = 9964 newtons
Stress	1 ton per square inch = 15.44 newtons per square millimetre
Speed	1 mile per hour = 1.609 kilometres per hour

References to the Brunel papers in the Library of the University of Bristol are given as follows.

CB	Calculation book (with date)
CSBT	Clifton Suspension Bridge Trust minute book (with date)
ESN	Eastern Steam Navigation Company books (with date)
NF	Notebook of facts (with date)
PLB	Private letter book (with date)
SB	Sketch-book (with number or date)

References to papers with British Rail are given as follows.

BD, 1840	Brunel's drawings at Paddington relating to Clifton Suspension Bridge
BS, 1840	Brunel's specification at Paddington relating to Clifton Suspension Bridge
BTCA	Notebook of facts (1841) with British Transport Commission Archives (with reference number: 1 = Rail/1149/11, 2 = Rail/1149/12, 3 = Rail/1149/13).

The numbers in the text refer to the following notes.

1. Sir Marc Isambard Brunel was known in his lifetime as Sir Isambard, but modern biographers and engineers have usually referred to him as Sir Marc to distinguish him from his son, Isambard Kingdom Brunel. This practice, although regretted by the family (Gladwyn, 1971), has been followed throughout this book.

2. For assessments of Brunel as a person, see his grand-daughter's book (Noble, 1938) and a recent paper by her daughter (Gladwyn, 1971).

3. For general information about Brunel's career reliance has been placed largely on the biographies by L. T. C. Rolt (1957) and by I. Brunel (1870). On Brunel's involvement in the Bristol Docks, see *I. K. Brunel and the Port of Bristol* (Buchanan, 1969).

4. The early volumes of the private letter books (PLBs) ran concurrently, so there is some overlap between them. The earliest entry was for 1834, but the letters were only recorded systematically from January 1836. They are in a variety of handwriting, but it seems likely that they were primarily the responsibility of Joseph Bennett, who became Brunel's chief clerk in 1836.

5. In addition to the reservations mentioned in the text, one of the greatest drawbacks of the private letter books is that they give only one side of the correspondence—that from Brunel's office. For somebody who depended so much on personal contact with both his staff and his professional colleagues, even the record of Brunel's letters is inadequate to explain the full sequence of events.

6. There is little evidence that Brunel was a snob in the usual sense of the word, but he invariably observed the conventions in terms of how people should be addressed and he had a strict view of conduct becoming a gentleman.

7. PLB: Brunel to H. E. Count Pollon, representing the Sardinian Railway, 18 Nov., 1845. It appears that Brereton took his wife with him: see Brunel to Babbage, 10 June, 1845, where Brunel says of Brereton, "he has had more experience than any man I have" on tunnel work.

8. PLB: Brunel to Gravatt, 3 Dec., 1839. There is no hint of the point at issue. The author is reluctant to identify this Gravatt with the rather lovable old engineer described by Rolt (1957), but internal evidence in the correspondence makes the identification virtually certain.

9. PLB: Brunel to Gravatt, 23 July, 1840. Headed significantly, "My dear Gravatt . . ." with "My dear" crossed out.

10. PLB: Brunel to Gravatt, 15 June, 1841. With a reference to "my past conduct towards you for nearly 15 ys."

11. PLB: Brunel to Bardham, 19 July, 1841, and 22 July, 1841, in which he offered his own resignation—"but I should still decline to defend myself—as I cannot admit that my character requires it".

12. PLB: Brunel to Marchant, 26 Mar., 1847. He wrote to Hulme the same day, reminding him that his assistants are men in authority. On 29 March, 1847, Brunel wrote to Marchant again, hinting that "being a relative" would not prevent him from being dismissed if he thwarted Brunel's intentions. See also 16 Feb., 1852, and 1 Dec., 1852.

13. PLB: Brunel to C. J. Darley, who appears to have been an assistant on a project at Bullo Pill in Gloucestershire, but he might have been a contractor; 28 Sept., 1857.

14. PLB: Brunel to C. Richardson, 14 Sept., 1858. Richardson accepted the invitation after some hesitation and was still in the post at the time of Brunel's death.

15. PLB: Brunel to Froude, 28 May, 1844. William Froude was one of a highly talented family, several members of which distinguished themselves in academic and ecclesiastical circles. Froude went on to establish the first naval testing tanks for ship designs. It appears that on this occasion in 1844 he wanted time off to nurse his sick father.

16. PLB: Brunel to Froude, 22 July, 1845. But he assumes that Froude will be too "gentlemanly" to accept it.

17. PLB: Brunel to Froude, 7 Apr., 1847. Froude later advised Brunel on the laws of motion of ships during the construction of the *Great Eastern;* see *Dictionary of National Biography* entry for Froude.

18. PLB: Bennett to Bell, 20 Oct., 1853. Bell seems to have been such a loyal and valuable assistant that the author is reluctant to believe that the person dismissed two years later for drunken behaviour (Brunel to W. Fleming?, 27 Nov., 1855), and apparently of the same name, could be he. But in the absence of any further information about Bell's career, the identification remains a puzzle.

19. PLB: Brunel to Owen, 16 Jan., 1836. It would rise to a maximum of £300 and, as with all such appointments by Brunel, carried the condition of instant dismissal.

20. PLB: e.g. Brunel to Bampton, 10 June, 1854, refusing to join a "Devon and Cornwall Society of Engineers".

21. PLB: see Brunel to G. Hudson, 4 Sept., 1844, offering to withdraw from any engagement "in favour of our Respected Grandfather G. Stephenson."

22. PLB: Brunel to W. G. Owen, 28 Feb., 1852. "Mr. Airy is a man who of course can thoroughly understand and appreciate such works—being a first rate mechanic as well as mathematician."

23. Brunel's difficulties have been so much emphasised by other writers that it is worth noting, for comparison, the rate of progress made on the Bletchingley railway tunnel in 1840. This was in dry swelling blue Weald Clay, and Simms (1844) records that a 12 foot length took twelve shifts to dig and 4½ days to brick-line.

24. These assessors, particularly Seaward, have been adversely criticised by Brunel's biographers, Rolt (1957) and Pudney (1974), but it is clear that Gilbert, who was an able applied mathematician, had a good influence on Brunel's design. From the company's standpoint, with Telford now as a competitor, it was essential that both assessors should not be directly associated with suspension bridge work, and Gilbert himself suggested Seaward, both for his mathematical knowledge and for his practical experience with ironwork.

25. The principal dimensions of the present bridge are: total span (centre to centre of piers) 702 feet 3 inches, length of suspended roadway 636 feet, width of roadway (centre to centre of chains) 20 feet, overall width (including pathways) 31 feet.

26. Brunel does not appear to have realised, once he adopted this system of staying the bridge, that he could have made it much more effective by 'damping' any motion of the dead weight by friction. Restraining cables with this sort of damped dead weight were used to stay the towers of the Forth suspension bridge during erection.

27. There is evidence on an existing drawing that, to investigate the condition of the Leigh Woods rock, Brunel drove a horizontal tunnel 230 feet into the hillside from a point just north of the abutment and some 80 feet below the level of the top of the abutment. This would have enabled him to study the rock right back to the anchorage region.

28. The 1840 contract specifications and drawings for the ironwork to be made by Messrs Carne and Vivian have recently been found in the possession of British Railways at Paddington.

29. Prior to the discovery of the 1840 ironwork specifications it was difficult to understand the phrase "before the holes are bored", but it now seems probable that what was meant was "before the holes are opened out to their full diameter". A clause to this effect appears in Brunel's specification. Such a step would have reduced the residual stresses (now deemed advantageous from a fatigue standpoint) put into the lugs by proof testing.

30. The scheme was no doubt the result of a desire to do everything possible to ensure that both chains should share the load equally; by the time he came to design the Hungerford bridge, Brunel had evidently decided that such mechanical elaboration was unnecessary.

31. During the Second World War, when the Menai bridge was reconstructed, the anchorage tunnels for the new steel chains were similarly filled with concrete.

32. Simms (1838) and Bourne (1846) indicate that the brickwork at the Wharncliffe Viaduct "upon striking the centering" followed it by "from 1¾ to 2¾ inches, averaging 2 inches".

33. In his calculation book *(CB, 1837)*, under the title "Brent Viaduct" Brunel deals with both the approach (Hanwell embankment) and the main Wharncliffe arches of the crossing over the River Brent. MacDermott (1927) also refers to a Brent viaduct in Devon, only the piers of which still remain and which is not far from a lovely Brunel viaduct at Rattery, which he also mentions.

34. Vignoles (1849) was rather less enthusiastic, but at least he thought it had the merit of "preventing the thrust coming upon the top of the wall". Neither Brunel nor Vignoles seems to have realised that any superimposed loads on the Lovelace roof, as from snow or wind, would have resulted in

some lateral thrust on the walls.

35. The origin of this reference is obscure.

36. Neither coefficient is correct. Both should be 0.0321.

37. The first iron Cunarder, built in 1855.

38. Slip, as used by Brunel, is defined by the equation

$$\text{slip} = (P-A)/P$$

where P is the pitch of the propeller blade and A is the actual distance advanced by the propeller in making one revolution in driving the ship.

39. That is, with the propeller operating in calm water without the disturbing influence of the hull ahead of it.

40. Moseley (1801–72) was Professor of Natural Philosophy and Astronomy at King's College, London, 1831–44, H.M. Inspector of Schools, 1844–53, and Canon of Bristol Cathedral, 1853–72.

41. J. H. Cotterill (1865) based his first attempt at establishing a principle of least work on Moseley's principle of least resistance.

42. It is believed that Bresse was the first to publish the use of symmetry and antisymmetry in structural analysis (Timoshenko, 1953).

43. It is interesting that, in his book *Economics of Construction in Relation to Framed Structures* (1873), R. H. Bow, originator of "Bow's notation", misrepresents the main span of Brunel's Chepstow bridge as a truss.

44. Moseley (1843) included energy methods for calculating the deflection of beams and, perhaps for the first time, showed that $dU/d\Delta = F$, where U is strain energy and Δ is deflection in the line of action of F, the force causing deflection. Moreover, he implied that $dU/dF = \Delta$.

45. W. Pole, successor to Vignoles as Professor of Civil Engineering at University College, London, did much to publicise the contents of Moseley's book, especially in respect of the strength of beams. He had collaborated with Moseley in experiments on the performance of a Cornish steam engine using the latter's form of indicator (Moseley, 1842; Pole, 1852).

46. During erection of the Britannia bridge an ingenious procedure was adopted whereby prestressing was introduced to equalize bending moments due to self-weight at mid-span and support points and thus derive the maximum possible benefit from structural continuity. W. Pole, a disciple of Moseley, was concerned with that matter. The novelty of the structure attracted a great deal of attention on the Continent, notably from Clapeyron and Jourawski (Timoshenko, 1953; Charlton, 1976), but Todhunter and Pearson (1886, 1893) appear to have misunderstood the concept. It is open to speculation whether Brunel's work on continuous beams with Bell bore any relation to the prestressing idea.

References

See also references to Brunel's unpublished papers listed at the start of the Notes section

Arkell, W. J. (1933). *The Jurassic System in Great Britain.* Oxford University Press.

Baker, W. A. (1965). *From Paddle Steamer to Nuclear Ship.* Watts, London.

Barlow, P. (1817). *An essay on the Strength and Stress of Timber.* Taylor, London.

Barlow, W. H. (1846). On the Existence (Practically) of the Line of Equal Horizontal Thrust in Arches, and the Mode of Determining it by Geometrical Construction. *Minut. Proc. Instn Civ. Engrs,* **5**, 162–182.

Barlow, W. H. (1867). Description of the Clifton Suspension Bridge. *Minut. Proc. Instn Civ. Engrs,* **26**, 243–257.

Beamish, R. (1862). *Memoir of the Life of Sir Marc Isambard Brunel.* Longmans, London.

Beaufoy, M. (1834). *Nautical and Hydraulic Experiments.* Beaufoy, London.

Belidor, B. F. (1739). *La Science des Ingénieurs.* Janbert, Paris.

Bell, W. (1871). On the Stresses of Rigid Arches, Continuous Beams, and Curved Structures. *Minut. Proc. Instn Civ. Engrs,* **33**, 58–165.

Berridge, P. S. A. (1969). *The Girder Bridge.* Maxwell, London.

Birkinshaw, J. (1820). *Manufacturing and Construction of a Wrought or Malleable Iron Railroad or Way.* British Patent 4503.

Booth, L. G. (1971a). The Development of Laminated Timber Arch Structures in Bavaria, France and England in the Early Nineteenth Century. *J. Inst. Wood Sci.,* **5**, 3–16.

Booth, L. G. (1971b). Laminated Timber Arch Railway Bridges in England and Scotland. *Trans. Newcomen Soc.,* **44**, 1–23.

Bourne, J. (1852). *A Treatise on the Screw Propeller.* Longmans, London.

Bourne, J. C. (1846). *The History and Description of the Great Western Railway.* Bogue, London.

Bow, R. H. (1873). *Economics of Construction in Relation to Framed Structures.* Spon, London.

Brees, S. C. (1837). *Railway Practice.* Williams, London.

Brereton, R. P. (1861). Description of the Centre Pier of the Saltash Bridge on the Cornwall Railway, and of the Means Employed for its Construction. *Minut. Proc. Instn Civ. Engrs,* **21**, 268–276.

Brereton, R. P. (1870). Discussion, On the Strength of Iron and Steel, and on the design of Parts of Structures which Consist of those Materials, G. Berkley. *Minut. Proc. Instn Civ. Engrs,* **30**, 270–272.

Brereton, R. P. (1871). Discussion, On the Theory and Details of Construction of Metal and Timber Arches, J. Gaudard. *Minut. Proc. Instn Civ. Engrs,* **31**, 155–162.

Brereton, R. P. (1881). Discussion, The Resistance of Viaducts to Sudden Gusts of Wind, J. Gaudard. *Minut. Proc. Instn Civ. Engrs,* **69**, 198–200.

Brunel, I. (1870). *The Life of Isambard Kingdom Brunel, Civil Engineer.* Longmans, London.

Brunel, I. K. (1846). Discussion, On the Existence (Practically) of the Line of Equal Horizontal Thrust in Arches, and the Mode of Determining it by Geometrical Construction, W. H. Barlow. *Minut. Proc. Instn Civ. Engrs,* **5**, 173, 175–177, 180.

Brunel, I. K. (1847). Discussion, On the Law which Governs the Discharge of Elastic Fluids under Pressure, through Short Tubes or Orifices, W. Froude. *Minut. Proc. Instn Civ. Engrs,* **6**, 386–387, 390–392.

Brunel, I. K. (1849a). Evidence in Report of the Commissioners appointed to Inquire into the Application of Iron to Railway Structures. HMSO, London.

Brunel, I. K. (1849b). Discussion, On the Construction of the Collar Roof, with Arched Trusses of Bent Timber, at East Horsley Park, Earl of Lovelace. *Minut. Proc. Instn Civ. Engrs,* **8**, 285–286.

Brunel, I. K. (1849c). Discussion, Description of Sir George Cayley's Hot-Air Engine, W. W. Poingdestre. *Minut. Proc. Instn Civ. Engrs,* **9**, 197–203.

Brunel, I. K. (1851). Discussion, An Investigation of the Strains upon the Diagonals of Lattice Beams, with the Resulting Formulae, W. T. Doyne and W. B. Blood. *Minut. Proc. Instn Civ. Engrs,* **11**, 14.

Brunel, I. K. (1853a). Discussion, On the Nature and Properties of Timber, with Descriptive Particulars of Several Methods, now in Use, for its Preservation from Decay, H. P. Burt. *Minut. Proc. Instn Civ. Engrs,* **12**, 233–234.

Brunel, I. K. (1853b). Discussion, On the Use of Heated Air as a Motive Power, B. Cheverton. *Minut. Proc. Instn Civ. Engrs,* **12**, 349–351.

Brunel, Sir Marc (1818). *Forming tunnels or drifts under ground.* British Patent 4204.

Buchanan, R. A. (1969). I. K. Brunel and the Port of Bristol. *Trans. Newcomen Soc.,* **42**, 41–56.

Buchanan, R. A. (1971). *Nineteenth Century Engineers in the Port of Bristol.* Historical Association, Bristol.

Burt, H. P. (1853). On the Nature and Properties of Timber, with Descriptive Particulars of Several Methods, now in Use, for its Preservation from Decay. *Minut. Proc. Instn Civ. Engrs,* **12**, 206–243.

Chalmers, J. B. (1881). *Graphical Determination of Forces in Engineering Structures.* Macmillan, London.

Chapman, F. H. (1775). *Tractat om Skepps-Biggeriet.* Stockholm.

Charlton, T. M. (1976). Contributions to the Science of Bridge Building in the Nineteenth Century by Henry Moseley, Hon.LL.D., F.R.S., and William Pole, D.Mus., F.R.S. *Notes Rec. R. Soc. Lond.,* **30**, 1960.

Cheverton, B. (1853). On the Use of Heated Air as a Motive Power. *Minut. Proc. Instn Civ. Engrs,* **12**, 312–351.

Clark, E. (1850). *Britannia and Conway Tubular Bridges.* Longmans, London.

Clements, P. (1970). *Marc Isambard Brunel.* Longmans, London.

Cook, S. S. (1948). Thermodynamics in the Making. *Trans. NE Cst Instn Engrs Shipbldrs,* **65**, 65–84.

Corlett, E. C. B. (1971). The Steamship Great Britain. *Trans. R. Instn Nav. Archit.,* **113**, 411–437.

Cotterill, J. H. (1865). On an Extension of the Dynamical Principle of Least Action. *Lond. Edinb. Dubl. Phil. Mag.,* **29**, 299–305.

Cubitt, J. (1852). A Description of the Newark Dyke Bridge on the Great Northern Railway. *Minut. Proc. Instn Civ. Engrs,* **12**, 601–612.

Dempsey, G. D. (1864). *Tubular and Other Iron Girder Bridges.* Virtue, London.

Doyne, W. T. (1851). Description of a Wrought-Iron Lattice Bridge, Constructed over the Line of the Rugby and Leamington Railway. *Minut. Proc. Instn Civ. Engrs,* **9**, 353–359.

Doyne, W. T. and Blood, W. B. (1851). An Investigation of the Strains upon the Diagonals of Lattice Beams, with the Resulting Formulae. *Minut. Proc. Instn Civ. Engrs,* **11**, 1–14.

Elgar, F. (1894). Fast Ocean Steamships. *Trans. Instn Nav. Archit.,* **35**, 20–40.

Ericsson, J. (1836). *Propeller applicable to Steam Navigation.* British Patent 7149.

Fairbairn, W. (1851). On Tubular Girder Bridges. *Minut. Proc. Instn Civ. Engrs,* **9**, 233–287.

Farr, G. (1963). *The Steamship Great Western.* Historical Association, Bristol.

Fletcher, L. E. (1855). Description of the Landore Viaduct, on the Line of the South Wales Railway. *Minut. Proc. Instn Civ. Engrs,* **14**, 492–506.

Froude, W. (1847). On the Law which Governs the Discharge of Elastic Fluids under Pressure, through Short Tubes or Orifices. *Minut. Proc. Instn Civ. Engrs,* **6**, 356–399.

Froude, W. (1861). On the Rolling of Ships. *Trans. Instn Nav. Archit.,* **2**, 180–227.

Gaudard, J. (1881). The Resistance of Viaducts to Sudden Gusts of Wind. *Minut. Proc. Instn Civ. Engrs,* **69**, 120–218.

Gladwyn, Lady Cynthia. (1971). The Isambard Brunels. *Proc. Instn Civ. Engrs,* **50**, 1–14.

Gordon, T. (1784). *Principles of Naval Architecture.* Evans, London.

Gray, M. (1869). The Steam Steering Apparatus Fitted in the Great Eastern. *Trans. Instn Nav. Archit.,* **10**, 101–118.

Green, B. (1839). On the Timber Viaducts now in Progress on the Newcastle and North Shields Railway. *Rep. Br. Ass. Advmt Sci.*, **7**, 150–152.

Gregory, O. (1825). *Mathematics for Practical Men.* Baldwin, Craddock and Joy, London.

Harland, Sir Edward (1894). Discussion, Fast Ocean Steamships, F. Elgar. *Trans. Instn Nav. Archit.*, **35**, 34–36.

Hawkshaw, J. (1853). Discussion, An Investigation of the Strains upon the Diagonals of Lattice Beams, with the Resulting Formulae, W. T. Doyne and W. B. Blood. *Minut. Proc. Instn Civ. Engrs*, **11**, 13–14.

Hemans, G. W. (1844). Description of a Wrought-Iron Lattice Bridge, Lately Erected on the Line of the Dublin and Drogheda Railway. *Minut. Proc. Instn Civ. Engrs*, **3**, 63–65.

Henderson, A. (1853). On the Speed and other Properties of Ocean Steamers, and the Measurement of Ships for Tonnage. *Minut. Proc. Instn Civ. Engrs*, **13**, 1–63.

Heyman, J. (1969). The Safety of Masonry Arches. *Int. J. Mech. Sci.*, **11**, 363–385.

Howard, T. (1845). *Rolling Iron Bars for Suspension Bridges and Other Purposes.* British Patent 10855.

Howard, T. (1849). Description of a Method of Rolling Bars for Suspension Bridges and Other Similar Purposes. *Minut. Proc. Instn Civ. Engrs*, **8**, 273–281.

Humber, W. (1870). *Treatise on Bridges*, 3rd edition. Lockwood, London.

Jeaffreson, J. S. (1864). *The Life of Robert Stephenson.* Longmans, London.

Jenkin, C. F. (1869). On the Practical Application of Reciprocal Figures to the Calculation of Strains on Framework. *Trans. R. Soc. Edinb.*, **25**, 441–463.

John, W. (1874). The Strength of Iron Ships. *Trans. Instn Nav. Archit.*, **15**, 74–93.

John, W. (1887). Atlantic Steamers. *Trans. Instn Nav. Archit.*, **27**, 147–178.

King, J. F. (1907). Structural Development in Merchant Ships. *Trans. Instn Nav. Archit.*, **49**, 287–299.

Lamé, G. (1852) *Leçons sur la Théorie Mathématique de l'Elasticité des Corps Solides.* Gauthier-Villars, Paris.

Landmann, I. (1815). *Treatise on Mines.* Bensley, London.

Lardner, D. (1838). On the Application of Steam to Long Voyages. *Rep. Br. Ass. Advmt Sci.*, **6**, 136.

Law, H. (1845). *Memoir of the Thames Tunnel.* John Weale, London.

Leeming, W. F. and Whitbread, J. E. (1974). Stress Investigation and Strengthening of the Royal Albert Bridge, Saltash. *Proc. Instn Civ. Engrs*, Part 1, **56**, 479–495.

Lindsay, W. S. (1876). *History of Merchant Shipping and Ancient Commerce.* Sampson Low, Marston, Low and Searle, London.

Longridge, J. A. (1855). *Construction or Manufacture of Guns and Artillery and of other Vessels Intended to Resist Great Pressure.* British Patent 1167.

Lovelace, Earl of (1849). On the Construction of the Collar Roof, with Arched Trusses of Bent Timber, at East Horsley Park. *Minut. Proc. Instn Civ. Engrs,* **8**, 282–286.

MacDermott, E. T. (1927). *History of the Great Western Railway.* G.W.R. Co., London.

Mackay, T. (1900). *The Life of Sir John Fowler.* Murray, London.

Matheson, E. (1873). *Works in Iron, Bridges and Roof Structures.* Spon, London.

Maxwell, J. C. (1864a). On Reciprocal Figures and Diagrams of Forces. *Lond. Edinb. Dubl. Phil. Mag.,* 4th series, **27**, 250–261.

Maxwell, J. C. (1864b). On the Calculation of the Equilibrium and Stiffness of Frames. *Lond. Edinb. Dubl. Phil. Mag.,* 4th series, **27**, 294–299.

Maxwell, J. C. (1869). On Reciprocal Figures, Frames and Diagrams of Forces. *Trans. R. Soc. Edinb.,* **26**, 1–40.

Moseley, H. (1833). On a New Principle in Statics, called the Principle of Least Pressures. *Lond. Edinb. Dubl. Phil. Mag.,* 3rd series, **3**, 285–431.

Moseley, H. (1835). On the Equilibrium of the Arch. *Trans. Camb. Phil. Soc.,* **5**, 293–313.

Moseley, H. (1842). Results of a Trial of the Constant Indicator upon the Cornish Engine at the East London Water-works. *Minut. Proc. Instn Civ. Engrs,* **2**, 102–121.

Moseley, H. (1843). *The Mechanical Principles of Engineering and Architecture.* Longmans, London.

Muller, J. (1770). *The Attack and Defence of Fortified Places.* Millan, London.

Murray, M. (1765). *A Treatise on Shipbuilding and Navigation.* Millan, London.

Navier, C. L. M. H. (1826). *Résumé des leçons données à l'Ecole des Ponts et Chaussées sur l'Application de la Mécanique a Establissement des Construction et des Machines.* Carilian-Gocury, Paris.

Nicholson, P. (1793). *The Carpenter's New Guide.* Taylor, London.

Nicholson, P. (1797). *The Carpenter's and Joiner's Assistant.* Taylor, London.

Noble, Lady Celia (1938). *The Brunels, Father and Son.* Cobden-Sanderson, London.

Nock, O. S. (1955). *The Railway Engineers.* Batsford, London.

Nock, O. S. (1962). *The G.W.R. in the 19th Century.* Ian Allan, London.

Owen, J. B. B. (1939). Edward's Single Span Arched Bridge at Pontypridd, 1755. *Proc. S. Wales Inst. Engrs,* **55**, 20–59.

Owen, J. B. B. (1965). *The Analysis of Light Structures.* Arnold, London.

Pippard, A. J. S. and Baker, Sir John (1962). *The Analysis of Engineering Structures,* 4th edition. Arnold, London.

Poingdestre, W. W. (1849). Description of Sir George Cayley's Hot-Air Engine. *Minut. Proc. Instn Civ. Engrs,* **9**, 194–203.

Pole, W. (1852). Discusion, On the Conversion of Heat into Mechanical Effect, C. W. Siemens. *Minut. Proc. Instn Civ. Engrs,* **12**, 593.

Pole, W. (1877). *The Life of Sir William Fairbairn.* Longmans, London.

Porter Goff, R. F. D. (1974). Brunel and the Design of the Clifton Suspension Bridge. *Proc. Instn Civ. Engrs,* Part 1, **56**, 303–321.

Pudney, J. (1974). *Brunel and his World.* Thames and Hudson, London.

Pugsley, Sir Alfred (1968). *The Theory of Suspension Bridges,* 2nd edition. Arnold, London.

Rankine, W. J. M. (1858). *Applied Mechanics.* Griffin, London and Glasgow.

Rennie, Sir John (1846). Address of Sir John Rennie, President, to the Institution of Civil Engineers. *Minut. Proc. Instn Civ. Engrs,* **5**, 19–122.

Rolt, L. T. C. (1957). *Isambard Kingdom Brunel.* Longmans, London.

Rouse, H. and Ince, S. (1957). *History of Hydraulics.* Iowa State University.

Royal Commission on Railway Gauges (1845). *Report.* HMSO, London.

Royal Commissioners Appointed to Inquire into the Application of Iron to Railway Structures (1849). *Report.* HMSO, London.

Royal Society (1860). Obituary notices of deceased fellows: Isambard Kingdom Brunel. *Proc. R. Soc.,* **10**, vii–xi.

Ruddock, E. C. (1974). Hollow Spandrels in Arch Bridges: A Historical Study. *J. Instn Struct. Engrs,* **52**, 281–293.

Russell, J. S. (1847). Discussion, On the Law which Governs the Discharge of Elastic Fluids under Pressure, Through Short Tubes or Orifices, W. Froude. *Minut. Proc. Instn Civ. Engrs,* **6**, 386–389.

Russell, J. S. (1860). The Wave Principle of Ship Construction, Parts I and II. *Trans. Instn Nav. Archit.,* **1**, 184–211.

Russell, J. S. (1861). The Wave Principle of Ship Construction, Part III. *Trans. Instn Nav. Archit.,* **2**, 230–245.

Russell, J. S. (1865). *The Modern Systems of Naval Architecture.* Day, London.

Sekon, G. A. (1895). *A History of the Great Western Railway.* Digby, Long, London.

Simms, F. W. (1838). *Public Works of Great Britain.* Weale, London.

Simms, F. W. (1844). *Practical Tunnelling.* Crosby Lockwood, London.

Sinclair, L. (1971). Discussion, The Steamship Great Britain, E. C. B. Corlett. *Trans. R. Instn Nav. Archit.,* **113**, 433.

Smeaton, J. (1759). An Experimental Inquiry Concerning the Natural Powers of Water and Wind to Turn Mills and Other Machines Depending on a Circular Motion. *Phil. Trans. R. Soc.,* **51**, Table VI, 165.

Smiles, S. (1859). *Self-Help: with Illustrations of Character and Conduct.* John Murray, London.

Smith, T. M. (1846). Account of the Pont-y-tu-prydd, over the River Tâfe, near Newbridge, in the County of Glamorgan. *Minut. Proc. Instn Civ. Engrs,* **5**, 474–477.

Snell, G. (1846). On the Stability of Arches, with Practical Methods for Determining, According to the Pressures to which they will be Subjected, the best Form of Section, or Variable Depth of Voussoir, for any Given Intrados or Extrados. *Minut. Proc. Instn Civ. Engrs,* **5**, 439–474.

Society for Promoting Christian Knowledge (1835). *The Book of English Trades.* Society for Promoting Christian Knowledge, London.

Stalkartt, M. (1781). *Naval Architecture, or the Rudiments and Rules of Ship-building*. Boydell, Dodsley, Sewell, London.

Stanhope, Earl C. (1790). *A Method of Constructing Ships and Vessels, and of Moving and Conducting Them with Great Velocity Without the Help of Sails, and also Moving and Conducting Them against Wind, Waves, Current, or Tide, or against the Power of Them all United.* British Patent 1732.

Stephenson, R. L. (1846). Discussion, On the Existence (Practically) of the Line of Equal Horizontal Thrust in Arches, and the Mode of Determining it by Geometrical Construction, W. H. Barlow. *Minut. Proc. Instn Civ. Engrs,* **5**, 174–177.

Straub, H. (1952). *A History of Civil Engineering.* Translation by E. Rockwell of 1949 edition. Leonard Hill, London.

Timoshenko, S. P. (1953). *History of Strength of Materials.* McGraw-Hill, London.

Todhunter, I. and Pearson, K. (1886, 1893). *A History of the Theory of Elasticity and of the Strength of Materials,* volumes 1 and 2. Cambridge University Press.

Tredgold, T. (1820). *Elementary Principles of Carpentry.* Taylor, London.

Vignoles, C. B. (1849). Discussion, On the Construction of the Collar Roof, with Arched Trusses of Bent Timber, at East Horsley Park, Earl of Lovelace. *Minut. Proc. Instn Civ. Engrs,* **8**, 286.

Whitley, H. S. B. (1931). Timber Viaducts in South Devon and Cornwall, G.W.R. *Rly Engr,* **52**, 385–392.

Index